Dark
and
Magical Places

ALSO BY CHRISTOPHER KEMP

The Lost Species:
Great Expeditions in the Collections of
Natural History Museums

Floating Gold:
A Natural (and Unnatural) History of Ambergris

Dark *and* Magical Places

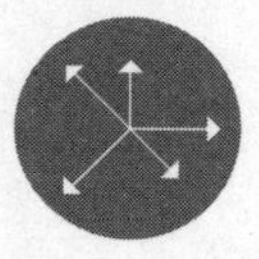

The Neuroscience of Navigation

CHRISTOPHER KEMP

W. W. NORTON & COMPANY
Independent Publishers Since 1923

Illustrations by Nate Kriepp

Printed in the United States of America
First Edition

Manufacturing by Lake Book Manufacturing
Book design by Chris Welch
Production manager: Anna Oler

ISBN 978-1-324-00538-4

W. W. Norton & Company, Inc., 500 Fifth Avenue, New York, N.Y. 10110
www.wwnorton.com

W. W. Norton & Company Ltd., 15 Carlisle Street, London W1D 3BS

1 2 3 4 5 6 7 8 9 0

For my family

(and anyone who has ever given me directions)

Contents

Author's Note

This is not a textbook. There are good textbooks out there on the subject of navigation, written by neuroscientists. I have used some of them in the writing of this book, like *Why People Get Lost*, by Paul A. Dudchenko, and *Human Spatial Navigation* by Ekstrom, Spiers, Bohbot, and Rosenbaum. There are others too. I relied even more heavily on the scientific literature, which is bursting at the seams with articles, reports, reviews, histories, and commentaries. With an ocean of data before me, I merely dipped my toe in it. Every month, a hundred or more new scientific papers are published, each describing a different aspect of the neural pathways that underpin navigation—a little piece of the puzzle. In most instances, I have provided a single scientific reference, or maybe two, to support a factual statement even though there might be hundreds of references in the literature that illustrate the same point. But I never intended to be exhaustive. This book is an attempt to understand the shortcomings of my own brain. Perhaps you see a little bit of yourself in here, or you recognize someone—a spouse, or a parent, or a friend—you've patiently guided through a grocery store.

Remember: not all those who wander are lost.

But some of us are.

CK 2021

The brain is a world consisting of a number of unexplored continents and great stretches of unknown territory.

—Santiago Ramón y Cajal (1852–1934)

I can't say as ever I was lost, but I was bewildered once for three days.

—Daniel Boone

Dad, I think you're going the wrong way.

—Izzy Kemp, age 7

Figure 1: Brain Regions Involved in Navigation

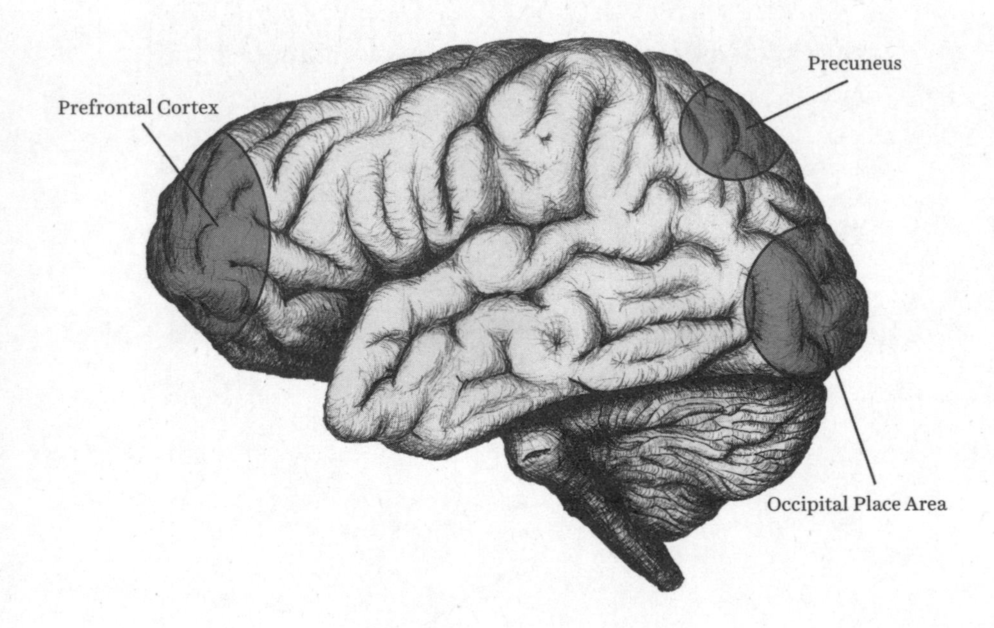

Figure 2: Internal Brain Regions Involved in Navigation

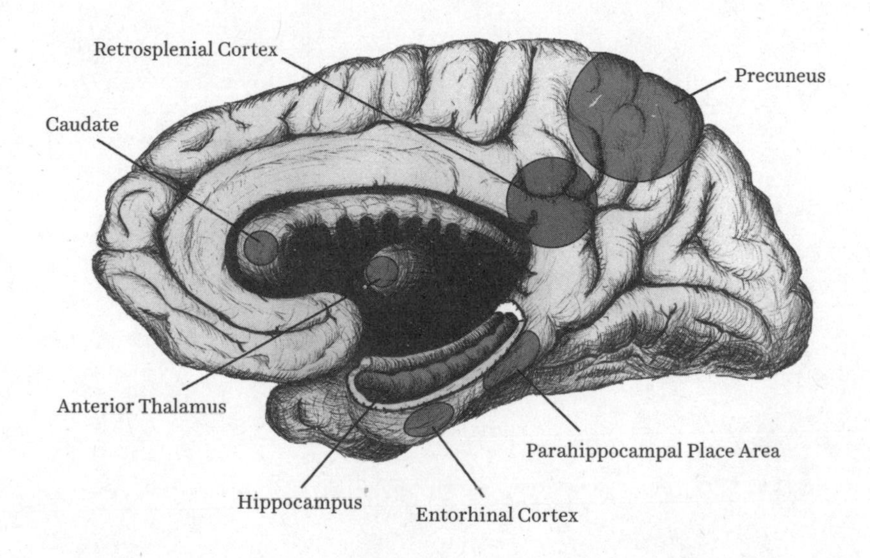

Dark
and
Magical
Places

Chapter One

Where Is Amanda Eller?

This is how it begins: in the forest, on an otherwise normal Wednesday afternoon, with sunlight filtering through the trees. Amanda Eller is hiking the trail. In the green shadows of the forest canopy, industrious little birds dart between the swaying stands of bamboo, hunting for grubs. As Eller walks farther into the woods, the day begins to get hotter. After a mile or so of walking, she's tired. She takes a few steps off the trail, lifting her feet carefully over a riot of ferns, and lies down on a fallen tree trunk. For a few quiet moments she rests, lying on her back with her eyes closed, face turned up toward the sky. Cartoon clouds drift across the gaps in the canopy. But when she stands up again and starts to head back toward the trail, she can't find it. Like an illusion, the trail has disappeared. Amanda Eller is lost.

At first, she does what any of us would do. She spins around trying to relocate the trail. All she sees are trees: tall straight trees with fluted trunks; saplings; spindly stunted trees growing in the shadows; a half-fallen tree resting at an odd angle against another canted tree, like a lambda λ in the forest; dead fallen trees that are slowly becoming part of the forest floor again; a tree with a large, rounded burl. Trying to detect something distinctive in their sameness, Eller searches for the trail all afternoon until it gets dark.

It's May 8, 2019. Eller is standing in the Makawao Forest Reserve in Hawaii. Located on the northwest slopes of the volcano Haleakalā, the reserve covers around 2,000 acres of Maui's rugged interior. It's a forest surrounded by even more forest. A wilderness within a wilderness. On Google Earth, it resembles a perfectly ripe avocado—if every avocado had a cartoonish flotilla of wispy clouds drifting above it. Her situation is not trivial. The forest is dense: completely impenetrable in places, steep and difficult, crisscrossed with hidden ravines, choked with ferns, and draped with vines. Eller had arrived at the reserve around 10:30am, parking her SUV at the trailhead. It was quiet. She bent over to place her car key beneath the front wheel of the car. The thirty-five-year-old yoga teacher and physical therapist had only intended to take a quick three-mile hike along a familiar trail.

Now, standing beneath the moss-green canopy, Eller closes her eyes and takes a slow, deep breath to stay calm. She studies the pale clumps of lichen growing on the nearby trunks in irregular shapes like the maps of unfamiliar countries. *Which way did I come from?*

But she is lost in a multitude of trees. By midafternoon, Eller has spent several hours searching for the trail. The sun is now high overhead, hanging above the forest. *This is so silly*, she thinks. Soon, someone will come ambling through the woods along the trail, voices will drift to her, and she can shout for help. Hikers will lead her through the tangle of undergrowth and the ferns, and back to the safety of the path. But no one comes. For a while, Eller follows a few narrow winding trails—eventually, she realizes they're paths made by wild boars as they shoulder through the undergrowth.

The birds trill and chirp in the trees, flitting from one place to another, as industrious as always. Somehow, at the same time, Eller's surroundings have both collapsed inward, to the size of a room, and expanded infinitely outward. She could be anywhere. Standing in the stillness of the trees, she suddenly hears everything, all the quiet, endless, subtle sounds of the woods.

Meanwhile, the sun continues on its arc overhead, slowly disappearing into the trees. It's as if the birds are singing it down. Then, it's dark. If she looks up through the swaying canopy, she can see a few stars—navigation tools, for someone maybe, but not for Eller. Eventually, her cell phone begins to vibrate, pinging with text messages from concerned friends. But she doesn't hear it in the woods. Wearing only a tank top and yoga pants, Eller has left it, along with her water bottle and wallet, in her car. She is unprepared. She had only planned to unplug for a while. So, heart hammering in her chest like a trapped bird, Eller sits at the base of a tree and waits for the first night to pass. The forest has swallowed her up.

By this point, perhaps some of you are already judging Eller. But for almost any of us, her situation would be terrifying. Recently, on a camping trip to northern Michigan, my seven-year-old, Rowan, woke me at 4am needing the bathroom. We crawled out of the tent together and stood beneath the wet roar of the woods in complete, oil-black darkness. A storm was forecast to arrive later that morning. The branches of the trees were bouncing madly in the wind, half-seen. I understood in a very small and incomplete way how mind-stoppingly terrifying it would be to find oneself disoriented and without shelter at 4am in the darkness of the woods.

Before she stepped off the trail and disappeared, Eller had been making her way through an unremarkable day. It was a very Wednesday kind of Wednesday. Suddenly, that had all changed. In one moment, her possible futures had begun to look very different—although she couldn't have known that yet. After all, if you set up the same scenario and run it again and again, in almost every other version she stands up, swings her legs back onto the ground, refreshed, and walks back to the trail. At the parking lot—a little thirsty now, and already thinking about lunch—she opens her car door, climbs inside, and glances quickly at her cell phone. No calls.

Almost every time, that's what happens. But not *every* time.

And definitely not this time.

•

How can someone take a few steps off a well-marked trail in a busy forest reserve and disappear?

I read about Eller's 2019 disappearance around the time it happened and was fascinated by the story. I followed along at home as it developed, reading updates every day. The story stayed with me for a long time afterward. It took root like a delicate sapling. Its leaves unfurled in my mind. I've been lost in the woods like Eller, if only for a few panicked moments.

In fact, I'm permanently lost. I have no sense of direction. If I were blindfolded and taken just a few blocks from my house in Grand Rapids, Michigan, where I've lived for more than ten years, I'd be as lost as if I'd suddenly teleported to the outskirts of Reykjavik. It takes me only a few moments to become disoriented. Crowded places. Darkened woods. Theme parks. Cities. My own street. Megastores. Stairwells. Little one-lane English villages in the rain. Everywhere, at night. I'm lost all the time.

In unfamiliar buildings, I begin to unravel. Once, I was trying to leave a hospital floor and I somehow walked deep into the building's innards instead. Suddenly, I was standing in a hot dark chamber surrounded by clanking pipes and half-filled buckets. My doctor's office is a maze-like warren of corridors that seem built to confuse. Malls, especially, are over-lit masterpieces of panic. A multistory parking lot, in its concrete brutalist glory, is a Soviet prison.

Art museums, however, are beautiful exceptions. The Metropolitan Museum of Art in New York is one of a handful of places in the world I am happy to be lost. I wander through its spacious galleries, bathed in its buttery light, propelled by a quiet and contented bewilderment.

Not long ago, I spent a week in a vast, sprawling Mexican resort. On one side, the ocean was a glimmering, nonnegotiable border. Inland, the resort was a green network of undulating paths—like a golf course without any holes. There were swimming pools everywhere. It was like being trapped in a David Hockney painting. I was lost for a week. One hot afternoon, lost yet again on my way back to our room, I realized that, earlier that day, I had used an iguana as a landmark—and it had moved.

On occasion—more often than I'd really like to admit—I drive past my own home. Maps help, but I'm still error-prone when I use them. I can learn the route to a particular location: to my dentist's office, or the airport, or my child's school. But if I don't take the route for a month or so, it slowly begins to decay, and then eventually disappears altogether. On a scale of 1 to 10, with 1 being very, very bad, I give my spatial skills a score of 1. And that's important.

Research has shown that when we're asked to rate our navigation abilities, our self-assessment is pretty accurate. In other words, we know if we're bad at it. My failures extend to all spatial tasks. I cannot mentally rotate an object or imagine the locations of the different rooms in my house relative to one another. Not only can I not read a map, I can't even refold one into an orderly rectangle. Jigsaw puzzles, crowded beaches, combination locks: all horrors. Origami is insufferable. A Rubik's cube is a deep and lasting humiliation. During a visit to the doctor's office, when I pull on my blue gown, I'm unable to tie a bow in the strings at the back of my neck. Driving into an unfamiliar city invokes sweaty, dry-throated, full-bodied, hyperventilating, white-knuckled, existential dread. Doom. Baroque panic. Gargoyles circle overhead.

Somehow, I managed to reach my early forties before I realized how incapable I am of orienting myself in space. It was a revelation. I used to just think that everyone was like me. But they're not.

My wife, Emeline, is an effortless and intuitive navigator. Her experience of the physical world is so different from mine that at times I've thought it was a brilliant and complicated lie. She might as well be telling me that she can talk with animals. But she proves her superior navigation skills to me all the time. Wherever she goes, she carries a complex and fully functioning map with her, stored deep within her brain. This inner map is richly detailed and informative. It bristles with spatial data. Neuroscientists have a term for it: a cognitive map. We can travel to a city we haven't visited for more than a decade and she will readily access her internal spatial map. In unfamiliar environments, she quickly begins to construct a new map—extending its borders, filling in its unknown margins.

How does she do this? Are our brains different, or do we just use them differently? And can I change how I use mine? This book is an attempt to try to understand the differences. To my wife, my never-ending state of lostness is a rich source of frustration. It doesn't make sense to her, to someone equipped with a well-functioning and constantly unfurling internal map of her surroundings, that I can be lost two blocks from home. At times, Emeline will suggest that I'm lost because I'm simply not paying attention to where I'm going. To me, this is like telling someone with color blindness that if he just studies the trees intently enough he will finally understand that they are green.

Rural places are difficult: rolling farmland, winding roads, bales of hay, birds perched on wires like musical notation. A tractor in a field. A scattering of crows. A silo. But cities are worse. Cities are dark and magical places. Hotels seem to disappear and reappear from one moment to the next, as if they're only accessible at certain times of the day. Streets disobey the physical laws of the universe. Entire city blocks seem to slide from one location to another. In a city—in *any* city—I am irretrievably lost. Smartphones have made them easier places to navigate. I have become a blinking blue cursor that drifts serenely along a street. But the dark magic remains. It prowls on the margins of the map like a distant storm, waiting for my phone battery to die.

So, when I first read Eller's story, I felt it: the heart-stopping moment of panic, the first wild seconds of rising fear, in the woods. The boar trails that led to nowhere. The cell phone, useless with its dark screen, back in the car.

It could have been me.

Chapter Two

Pink Seahorses

A normal adult human brain weighs about three pounds and contains around a hundred billion interconnected neurons. It is endlessly complicated—an organic supercomputer. Take just one of those neurons: it can receive information from as many as 10,000 other neurons and connect with another 10,000 neurons, forming a vast and interconnected network. It is an endlessly firing, electrical web. Until recently, the complex systems in the brain that control navigation have remained a mystery.

But that has now begun to change. Within the last several decades, neuroscientists have developed a much deeper and more detailed understanding of how the brain enables us to navigate the world—a fundamental aspect of human existence. The neural systems that underpin navigation are ancient: they're older than language, older than culture. They allowed us to disperse as a new species, radiating outward from the Rift Valley in Africa to eventually occupy and alter every corner of the world.

Neuroscientists have now identified and described populations of cells that perform very different and specific navigational functions—cells with names like place cells, grid cells, and head-direction cells. Certain brain regions work together to help us navigate, too: the hippocampus, the prefrontal cortex, the parahippocampal place area, the entorhinal and retrosplenial cortices, the caudate nucleus—all working in concert to provide us with a constantly unfolding map of the world and to help us navigate it. There are parts of the brain that have a special affinity for landmarks, and specialized neurons that respond only when their owner approaches a boundary in the environment, and brain regions that plan routes. But many mysteries still remain.

A 2019 study showed that human brain waves respond to changes in the Earth's magnetic field, which implies that we might subconsciously use its invisible force to navigate by—like eels, or migratory birds. We need all of these cells and systems and different brain regions to function properly together. When they don't, our ability to find our way quickly begins to fall apart.

Navigation is a whole-brain event. Perhaps you're not a chess grandmaster processing the complex layout of the board, visualizing your next seven moves, and simultaneously weighing a million different possibilities. Perhaps you struggle with long division. It doesn't matter: navigating a crowded IKEA store without getting lost is still an infinitely more complex neural task. Failures in the system occur partly because navigation is such a profoundly complicated neural process. Just look at the *Oxford English Dictionary* definition of it:

> **Navigation**: nævɪ'geɪʃn (mass noun) The process or activity of accurately ascertaining one's position and accurately planning and following a route.

In other words, navigation is not even a single thing. Even by the *OED*'s decidedly nonscientific definition, it's a combination of at least three very different functions: knowing where you are, planning a route through an environment, and then taking it. If we fail at any one of these steps, we fail to navigate. Each of these distinct components of navigation requires its own complex cascade of neural events to take place—a waltz between different brain regions. And it never stops. We're *always* navigating. Every time we open the front door and walk out to our parked car, or navigate the aisles of a grocery store, we're relying on our navigation systems to get us where we need to go. We even navigate within our own homes. For this, we rely on a small and powerful part of the brain called the hippocampus.

•

One day in the 1560s, Julius Caesar Arantius carefully hefted a fresh human brain onto his workbench. An anatomist at the University of Bologna, Arantius stepped back to survey the brain in the pearly light. It settled slowly under its own weight: a glistening pink dome of sinuous grooves and ridges, its exterior marked by a delicate riverine tracery of strawberry-red veins. Straightening, Arantius carefully placed his sharpened bone knife against the cortical surface of the brain and pushed downward evenly. The brain fell into two pieces, like a halved apple. Bending close to the cut surface, Arantius inspected the densely packed interior of the brain. He studied its intricate geography—the coastline of the cortex, with its numerous inlets and fjords, its inundations. Then, taking a deep breath, he hunched close to the brain and stuck his searching fingers into it.

Located deep in the center of the brain, Arantius found a small C-shaped structure. It sits almost atop the brain stem, insinuating itself between the other complex structures. From there, it sweeps backward, arching up and over itself in an elegant half-loop like a Nike swoosh—one hippocampus in each lobe of the brain. Working slowly, Arantius separated one of them from the surrounding tissues and laid it wetly on the workbench in front of him, intact. With its long curve—known as the *fornix*, Latin for "arch"—Arantius thought it resembled a seahorse, or a ram's horn, or perhaps a white silkworm. It's a beautiful object.

By 1564, when he published his first description of the structure in *De Humano Foetu Liber*, Arantius had finally settled on the name *hippocampus*, which means "seahorse," from the Greek: *hippos* for "horse" and *kampos* for "sea monster." If the hippocampus is dissected from a human brain and arranged upside-down on a workbench, it really does look a lot like a strange pale-pink seahorse—its fornix curling beneath it like the tapering tail of a comma. The name stuck: the seahorse. It was a remarkable discovery, a glimmer of light in the darkness.

For context, when Arantius described the hippocampus, people still firmly believed in the existence of witches. In England, Queen Elizabeth's parliament had recently passed an anti-witchcraft law: *An Act agaynst Conjuracions Inchantmentes and Witchecraftes.* Its first victim was Elizabeth Lowys, who was hanged that year for using her dark powers to kill three people. Most common physical ailments were treated with bloodletting. Crimes were punished with brutal torture. Solar eclipses were bird-silencing portents of doom. Alchemists were searching for an elusive chemical formula that would enable them to transform lead into gold. These were dark times. The same year, a hundred miles from Arantius, across the Apennines in Pisa, the astronomer Galileo Galilei was born. A scientific awakening was underway, but it had only just begun. Today, neuroscientists understand that the hippocampus plays a vital role in neuroplasticity, and in complex brain functions like learning and memory. It is central to spatial memory and navigation too. But remarkably, in 1564 and for most of the next four centuries, the function of the hippocampus remained a complete mystery.

It wasn't until the 1950s that neuroscientists finally began to understand what the hippocampus does. When they did, it was from the brain of a single person. Four hundred years of mystery ended because of the misfortune of a man called H.M. For decades, the world only knew him by his initials, in order to protect his privacy. Finally, when he died in 2008 at the age of eighty-two, we were told his full name: Henry Gustav Molaison.

When H.M. died, his brain was removed, fixed in formalin, frozen in gelatin, and then cut into 2401 slices on a microtome. Each coronal section—a perfect cross-section through the brain—is seventy microns thick, or about the thickness of a human hair. The process took a grueling fifty-three hours of nonstop cutting and was broadcast live via webcast. It's hypnotic to watch: a wide metal blade slides slowly, as regular as a metronome, across the surface of the frozen brain, encased in a frosted block and covered in a layer of ice. After each pass of the blade, neuroscientist Jacopo Annese retrieves a perfect brain section with a soft-bristled brush, depositing it in a container for long-term storage.

From the age of ten onward, H.M. had suffered from epileptic seizures. Slowly, as the years passed, the seizures grew more frequent and severe, until a normal life became impossible. On September 1, 1953, when H.M. was twenty-seven, a neurosurgeon named William Beecher Scoville decided to try something radical: a bilateral medial temporal lobe resection. It was an invasive and ethically questionable surgery. Scoville drilled two small holes in H.M.'s forehead while he sat conscious and upright in a chair, and then inserted surgical tools through the holes, guiding them into his brain like a game of *Operation*. Nudging the frontal lobes gently aside with a spatula, Scoville suctioned out almost all of H.M.'s hippocampus, along with some of the surrounding cortical brain regions. It did the trick. H.M.'s seizures were gone. But now he had a new problem to deal with.

"H.M.'s basic problem was that he was profoundly forgetful," says Larry Squire, a memory researcher at the University of California, San Diego. That's an understatement. After the surgery, H.M. remembered almost nothing. In a single moment, his long-term memory had been permanently erased—removed. The damage to H.M.'s brain was widespread, says Squire (who, in self-rating his spatial skills, gives himself a score of 8 or 9 out of 10). "It involved the hippocampus, the amygdala and all of the underlying cortex—the entorhinal cortex, and the perirhinal cortex, and the parahippocampal cortex." Before H.M.'s surgery, neuroscientists believed that memories were encoded via a process that engaged the whole brain. Clearly, that was not the case.

There are different kinds of memory, and each one relies on different brain structures. The shortest lived is sensory memory. It's fleeting. It degrades quickly—in perhaps less than a second—or is transferred to short-term memory. The information transferred to short-term memory lasts slightly longer, as long as twenty or thirty seconds. We employ the short-term store when we remember a series of digits like a phone number or perform a brief task. Current research suggests the prefrontal cortex is involved in short-term memory.

In his landmark 1890 text *The Principles of Psychology*, the American psychologist William James defined short-term memory—he called it *immediate memory*—as "belonging to the rearward portion of the present space of time, and not to the genuine past." Short-term memories either degrade and disappear, or they are encoded into permanent long-term memories via a process known as *long-term potentiation*, which takes place in the hippocampus. Long-term memory is more permanent. It can be further divided into episodic memories, which are personal autobiographic details about events, and semantic memory—the memory of facts. Long-term memory is a kind of deep storage. Or, as James put it: "An object which has been recollected is one which has been absent from consciousness altogether, and now revives anew. It is brought back, recalled, fished up, so to speak, from a reservoir in which, with countless other objects, it lay buried and lost from view."

In tests, H.M. could maintain a short conversation, and complete brief tasks. He could repeat a string of numbers back to researchers. In other words, his short-term memory was intact—his prefrontal cortex was mostly unaffected. But his long-term memory was profoundly affected by his surgery. His episodic memory was erased. He had no personal memories of his life, but he retained some of the semantic factual memories from before his surgery.

The exact process by which memories are made remains poorly understood. Consider the conversation we are having right now, Squire says when I reach him by phone: "Something is happening that's never happened before in either of our lives. There are things that are spatially specific and temporally specific. In all respects, this is unique."

This single unrepeatable moment activates brain regions involved in perception, in short-term memory, and in forming long-term memory—all parts of the brain that have never been activated together at the same time. With its network of connections to other brain regions, the hippocampus binds together all those different sites, Squire tells me. "Once we hang up the telephone and the episode is gone from our immediate perception, we have a way of re-invoking it because the hippocampus has held it all together," he says. Across time, the connectivity between those different areas of the brain can be reinforced, growing stronger until the assembly of these things—of time and space—coheres into a single representation. The specific details of the moment—that it is a Wednesday afternoon in March, and a truck is rumbling down my street in the rain—become permanently tied to one another. Eventually, the details can be invoked without the help of the hippocampus. Together, they have formed a memory in the cortex. "This gradual process of transferring responsibility from the hippocampus to the cortex is called consolidation."

The cortex—the storeroom of memory—is like a vast and limitless library. A 2016 study estimated the brain's memory capacity at a quadrillion bytes of data, or a petabyte—around ten times greater than previous estimates. For context, in April 2019, NASA released the first captured image of a black hole, located in the core of the supergiant elliptical galaxy Messier 87, in the constellation Virgo, 55 million light-years from Earth. The image was constructed from a gargantuan five petabytes of data—the memory capacity of just five human brains. After his 1953 surgery, the process of memory consolidation stopped for H.M. and his library of memories received no new information. No new books, as it were. In 1957, when doctors asked him what year it was, he told them it was 1953. Mysteries remain, says Squire. "We still don't really know how memory is stored in the long term," he says, "or what the mechanism is, or how the cells actually encode memory, or where they are."

In recent years, even this widely accepted model of long-term memory has been called into question. "That is the standard story that has been told for quite a long time, and I was one of the people who started to tell that story around thirty or forty years ago," says Lynn Nadel, a psychologist at the University of Arizona (an 8 or 9, who says he's "hopeless at spatial rotation problems"). "But the evidence does not support that story very well anymore," he says. "A lot of the memories that the hippocampus is involved in storing will end up being stored outside the hippocampus, but some remain. The ones that remain are the more detailed ones. The hippocampus seems to remain critical for highly detailed vivid memories, and for access to highly detailed vivid memories."

One thing is clear: the hippocampus is central to forming and storing memories, and therefore it's important in navigation too. "It's difficult to disentangle navigation from memory," says Robert Clark, a researcher at the University of California, San Diego. Navigating without a hippocampus—an impossibility—would be like walking without legs. Like many researchers, Clark (a confident 10) uses an animal model for his experiments—typically, rats. "I always look at these questions from the human perspective and then try to model my animal studies to get at what's interesting me in the human result," he says. In a typical animal study, researchers investigate the function of a specific brain structure by intentionally injuring it—creating what's known as a *lesion*—and then examining the effects. For instance, if Clark wants to understand the role of the hippocampus in navigation, he lesions that part of a rat's brain and then subjects the rat to a spatial task.

"If you make a lesion and it screws up navigation, that could be because the animal can no longer perform the computations necessary to accomplish navigation," he says. "But it also could be because in order to navigate successfully you have to be using memory."

Either way, without the ability to form new memories, navigation becomes completely impossible. Alongside his memory deficits, H.M. struggled to perform all kinds of spatial tasks. He wasn't alone, either.

In 1985, an Englishman named Clive Wearing contracted herpes simplex encephalitis, a rare neurological infection that leads to swelling in brain tissues. Eventually, doctors were able to control and treat the infection, but not before the swelling had caused significant brain damage. His hippocampus was destroyed. Before the infection, Wearing had been an accomplished musician and musicologist. He played the piano; he was a tenor singer. He arranged and conducted the complex polyphonic compositions of sixteenth-century Renaissance composer Orlande de Lassus (a contemporary of Julius Caesar Arantius). After his illness, Wearing's amnesia was so severe he could no longer function normally. Every few moments, his consciousness restarts again and again in an endless loop of forgetting and then beginning again—as his wife, Deborah, puts it, "An ever-repeating moment of first awakening." Sometimes, these moments of consciousness last as long as thirty seconds—the lifespan of a short-term memory. Usually they're shorter—ten seconds perhaps, or seven. He lives in a state of constant confusion.

A 1988 episode of the BBC series *The Mind* profiled Wearing and his profound memory loss. In one scene, he sits on a park bench in pale autumn sunlight. He's handsome, with a long, prominent nose and a square jawline. Deborah sits next to him, pale and pretty, with a nimbus of frizzy brown hair. As cars speed past on the street behind them, Deborah gently presses him, "Do you know how we got here?"

CLIVE: No.

DEBORAH: You don't remember sitting down?

CLIVE: No.

DEBORAH: I reckon we've been here about ten minutes at least.

CLIVE: Yeah, well, I've no knowledge of that. My eyesight has started working now, and all I've seen the whole time I've been seeing anything at all is that.

DEBORAH: And do you feel absolutely normal?

CLIVE: Not absolutely normal, no. I'm completely confused.

DEBORAH: Confused?

CLIVE: Yes, I've never eaten anything, never tasted anything, never touched anything, never smelled anything. By what rights do you assume you're alive?

DEBORAH: But you are.

CLIVE: Well, apparently, yes. But I'd like to know what the hell's been going on.

Ten seconds: cut. Begin again.

•

Before it was destroyed by a virus, Wearing's hippocampus—curved like two minuscule tapering commas—weighed just a fifth of an ounce. In other words, it weighed about the same as a standard sheet of printer paper. The industrious hub of his memory, it represented about one half of one percent of his brain's total weight, but it's difficult to overstate its disproportionate power when considering its diminutive size. "One can think of the hippocampus as containing specialized equipment—temporal cells, time cells, place cells—that together are the building blocks of a memory," says Squire. Space and time are the scaffold on which memories are built. And space and time are the raw materials of navigation. Essentially, navigation is an act of remembering, a seamless combination of sensory memory, and short-term and long-term memories spliced together, interpolated and intertwined with one another by the hippocampus and other related brain structures. Even so, our knowledge is provisional.

"On one level, we think we know a lot about the hippocampus," says Lynn Nadel, "and I can certainly say we do know a shitload about the hippocampus at this point. On another level, we don't have a clue about what it's doing. The pictures of the black hole [NASA] just showed us—on some level we sort of understand the universe because we're able to function in it, and make things work in certain ways, and make predictions. On another level we don't really have a clue, do we? Is it a 27-dimension string multiverse? I haven't got a clue. With respect to the hippocampus, it's kind of like that."

To neuroscientists like Nadel, the inner spaces of the brain present similar challenges to those astronomers confront when they peer into the distant parts of the universe. "We have an immense amount of detail for the hippocampus," he says. "We have a rough idea of what it's probably doing, but then when you back up and you ask '*Really*, what is it doing? What is this network doing? How do all these pieces work together?' we are, a little bit, at square one."

•

On any day, a search-and-rescue team is out there somewhere, combing the remote backwoods of New England, or the waterless scrub of Death Valley. A helicopter is flying low across the Scottish moors. Or the Australian outback. Or the bleak monochromatic Icelandic coastline: a white helicopter zipping over flat black rock. As you read this, a K9 handler is coaxing a team of eager dogs into a half-forgotten gully somewhere. The dogs plunge into the ravine, noisy, straining against their leashes, noses pushed to the ground.

It's a simple fact that people get lost all the time, and sometimes with disastrous consequences. Sometimes the stories are surreal. I have begun to collect them: a stack of strange-but-true tales on my desk. One morning in July 2013, hiker Geraldine Largay stepped off the Appalachian Trail in remote western Maine to relieve herself in the woods. She never found the trail again. As squirrels spiraled madly up and down nearby tree trunks, she texted her husband, who had been bringing her supplies along the trail:

> In somm trouble. Got off trail to go to br. Now lost. Can u call AMC to c if a trail maintainer can help me. Somewhere north of woods road. Xox.

But Largay was in a remote and thickly wooded part of Maine—and far from a cell phone tower. Her phone was useless. With no signal, the text message waited in limbo in her phone—and that's where it stayed. She'd bought a GPS device, but had left it behind in a motel room. She had a compass, but didn't know how to use it. Instead, she climbed uphill with her flip phone in her hand, searching for a cell phone signal. She wandered in circles. Her navigation systems failed her. In the state of Maine alone, where Largay disappeared, about two dozen hikers go missing each year from the trail. Most are found within forty-eight hours. But Largay was adrift in a green sea.

In January 2018, Tyler Batch had been driving along a quiet forest service road in the remote coastal Oregon backwoods, steering a borrowed Dodge Stratus over rutted tracks, when the car broke down. He abandoned it, jumper cables strewn across the front seat. Wearing a hoodie and logging boots, and carrying a gallon jug of water, the thirty-four-year-old walked into the woods. Rather than retrace his journey on foot—it would have meant a thirty-mile trek—he decided to take a shortcut through the forest instead. Batch wandered around the remote backwoods for the next seven days.

When his boots rubbed his feet raw, he walked barefoot through the woods. In desperation, he walked into creeks and let them roll him downhill like a boulder. "They zigzagged just like the roads," he said. "I've been in the woods most my life and that wasn't something I'd think would happen." Eventually, after a harrowing week in the woods, Batch crossed a road and was rescued by passing motorists.

In March 2017, a search party found Chilean tourist Maykool Coroseo Acuña after he'd spent nine days unable to find his way out of the thick Bolivian rainforest. He told rescuers that monkeys had kept him alive in the forest, throwing food down to him from the trees and leading him to water every day.

In July 2017, twenty-five-year-old Lisa Theris ran into the backwoods in Alabama when the two men she was with decided to ransack a hunter's campsite. It was an impulsive decision. She was gone for a month. Finally, when a woman drove past Theris at the side of a backcountry road, she thought she had seen a deer. Naked and covered with cuts and dirt, Theris had survived on berries and mushrooms and by drinking muddy water. Local police suspect she was high on methamphetamine when she ran into the woods, stripping off her clothes as she ran between the trees. "We are thinking they were all pretty much on drugs," Bullock County Sheriff Raymond Rodgers told the *Daily Mail*.

Inevitably, some of the stories end in disaster. It must be the loneliest death of all.

In October 2015—more than two years after her disappearance—Gerry Largay's remains were found by a forestry surveyor. She had been less than half a mile from the trail. Her camp was overgrown. A collapsed tent, half covered in leaves. A journal with moss growing across it. A map. A rosary. A flashlight that still worked. And a skull wrapped in a sleeping bag. Largay had survived for almost four weeks in the woods after she stepped off the trail. During the search, three canine search teams had come within a hundred yards of her camp. In her moss-covered journal, in a note dated August 6, more than two weeks after she disappeared, Largay had written:

> *When you find my body, please call my husband George and my daughter Kerry. It will be the greatest kindness for them to know that I am dead and where you found me—no matter how many years from now.*

And as if the terrifying experience of being lost isn't enough, the people who disappear in forests, canyons, deserts, and mountain passes must experience another indignity: they have to endure the disapproval of the crowd. People read the stories in newspapers or watch segments on the evening news and simply can't imagine taking a wrong turn in a pathless forest and getting lost. The message from popular culture is simple: intelligent and capable people don't get lost. Within a few days of her disappearance, the official search for missing hiker Amanda Eller was suspended. Volunteers stepped into the void left by the retreating officials and continued the search. In its wisdom, the crowd blamed her, asking: *Why had she left her cell phone and water bottle in her car? Who leaves a key under a tire?* Without thinking, Datch had taken an unknown shortcut through unmanaged forest instead of simply retracing his steps. Gerry Largay couldn't use a compass.

And what does it even mean to be lost? Every classification system has its own taxonomy, and being lost is no different. There are different degrees of lostness, outlined in a 1988 paper by Newcastle University geographer Jeremy Crampton (a 6 with a map). For instance, there is the hiker who knows she is lost. She is the known lost. She has a significant problem: she doesn't know where she is. But at least she knows that she's lost. She can try to backtrack to familiar terrain, or climb to higher ground, or use the contours of the landscape to navigate by, or use a compass if she has one, or try to pinpoint her position by the sun, or find a road, or follow a river downstream. She has many options. But there's another level of lostness, and it's much more dangerous. It's a different continent of lostness—another realm. This is the experience of the hiker who doesn't even know she is lost. She has stepped off the edge of her cognitive map, but she doesn't know it yet. Instead, she travels deeper into unknown territory, farther into unfamiliar terrain. She gets more lost by the minute. Two roads diverged in a wood and she has taken the one less traveled—but she didn't mean to take it. Crampton identified this hiker as the unknown lost. And this hiker, the unknown lost, is in serious trouble. Eventually, Eller realized she was lost. Until then, every effort to reach the trail had just taken her farther from it.

•

Just after the Second World War, a behavioral psychologist named Edward Tolman stood over a blocky wooden maze in his research lab at the University of California, Berkeley. To the west, across the San Francisco Bay, the sun was sinking into the ocean in flames. In the Berkeley Hills above campus, the chaparral and the sage scrub were disappearing into the gun-metal shadows of twilight. Balding and bespectacled, Tolman looked like a bank manager—if bank managers wore stained lab coats and held a rat in each hand. In old photos, his clipped brown mustache sits on his top lip like a little twin-peaked mountain. His owlish eyebrows are perpetually raised in matching quizzical arches. But he's a proto-hippie. A few years later, at the height of the McCarthy era, he would briefly lose his post at Berkeley after steadfastly refusing to sign a loyalty oath that stated he didn't belong to the Communist Party.

If there was one thing Edward Tolman believed in more than anything else, it was a maze. To Tolman, a well-designed maze had the power to reveal the brain's secrets like nothing else. He'd said as much a few years earlier, in an address to the American Psychological Association. Every important question that still remained in the field of psychology, Tolman told the gathered psychologists, could be answered by watching a rat making decisions in the confines of a maze. In other words, mazes were in. By then, they'd already been in for around half a century. A maze was a window into unseen cognitive processes, particularly those that govern spatial learning and memory. It was a way to peer into the black box of consciousness. And so: there were T-mazes, and X-mazes, and Y-mazes, and even intricate scaled-down replicas of the centuries-old Elizabethan hedge maze at Hampton Court in London—all with inquisitive mice scampering along their corridors instead of bewildered people. In an early maze study, one psychologist had even made a large outdoor maze for human subjects. He sent them stumbling through its alleys blindfolded, a bag of flour with holes in it tied to each person so that he could track their progress through the maze.

A note on mazes: they're still in. Modern scientists can monitor the electrical activity of a single neuron, but almost everything we know about navigation has been learned by studying animals as they navigate a maze. The arrival of virtual reality technology has only made mazes even more powerful and informative. Until now, using a maze to study the spatial abilities of human subjects has been impractical: what research institution could house a maze large enough for humans to navigate? Who has that many flour bags? But now human subjects can walk through a virtual maze—a scaled-up Hampton Court maze, or an entire simulated lifelike city, a virtual San Francisco or London—on a desktop computer, or by putting on a pair of virtual reality goggles.

More than a century after scientists first starting using them, mazes still allow researchers to place an animal in a complex environment and monitor all of its spatial decisions, testing its spatial memory. The possibilities are almost endless. The radial arm maze is one of the most widely used types of maze in studies of spatial learning. It was developed in 1976. The Morris water maze is another, born in 1981. The elevated plus maze was invented in 1984. At this moment, a bee expert is carefully loading a bumblebee into a scaled-down bee-sized version of a radial arm maze, its pathways radiating outward like the spokes of a wheel. In each arm of the maze, there is an artificial flower. Using this approach, researchers can test the effects of chronic insecticide exposure on the spatial memory of bees.

Other scientists are learning about cognition in cuttlefish species by watching them navigate between the clear walls of a saltwater-filled

T-maze. There are specially adapted mazes that bats can fly through, echolocating their way, as well as a binary tree maze that allows scientists to investigate how a solitary roving scout ant recalls the spatial location of food and communicates it to other ants in its colony. Caterpillar researchers are trying to understand exactly how a larval caterpillar manages to retain its spatial memories during the cellular upheaval of metamorphosis by watching it navigate a modified Y-maze, first as a larva and then again as an adult moth.

Botanists, too, are studying the neurobiology of different plant species with mazes, planting seedlings and measuring their growth through modified mazes under different environmental conditions. For a 2020 study investigating how cells migrate through tissues, researchers even built miniature mazes for cells to navigate. In fact, I just watched a short video of a *Dictyostelium discoideum* amoeba cell solving a scaled-down version of the Hampton Court maze—a cell rampaging down a narrow corridor. Videotaped from above, I know it's a single cell, but it looks like a drunk tourist.

Once, during a family visit to the Museum of Science and Industry in Chicago, I spent a harrowing period of time—I am told minutes—in something called a mirror maze. The maze is constructed of countless tessellated triangular mirrors. It is a nightmare, an endless nightmare. A torture chamber in downtown Chicago. Deep within the mind-altering mirror maze, every time you round another tight and darkened corner, you encounter yet another version of yourself. Sometimes, three or four of you will converge at the same point in space. You might see yourself in profile, three-quarter view, and head-on all at the same time. And all four of you will lurch together toward a mirror bathed in disorienting neon light, as if you're all running to try to catch the same ball. You see yourself in infinity. You see yourself growing confused and turning around in a panic so sudden that the sides of your coat fly into the air and seem to hang there forever.

Running past us, other families were clearly experiencing unbridled joy in the mirror maze. The infinity of it. But although I was with my family, I was on the edge of a total loss of self. I was disintegrating. I was not just lost, but I was also watching several versions of myself be lost at the same time. In the half-darkness, I saw my selves becoming scared. Reaching yet another mirrored dead end, I began to wonder how to raise an alarm in the maze. How long was it supposed to take us to complete the maze? Had we been in it for a day? Any sense of the direction we had come from, or the different parts of the maze we'd already visited, was gone. Time was altered in an unnerving way. It had unspooled. It was a humiliating experience.

In the end, my oldest son, Max, led me out of the maze. He took me gently by the hand and pulled me along, through the strangely Arctic neon light, to the exit. I don't know how he did it. He was six or seven at the time. But somehow, he knew where to go.

Back in Berkeley in 1945, one of Tolman's graduate students stepped forward and carefully placed a rat at the entrance to the maze that Tolman had built. Down below in the distance, the lights of North Beach and the Mission District glittered like another maze. At the time, a system of thought known as behaviorism was shaping the field of psychology. It was radical. Behaviorists argued that since consciousness is unobservable, it could be ignored. Instead, the complexity of cognition should be studied through the prism of behavior.

Nothing is true, the behaviorists said, unless it can be observed in behavior. Strict behaviorists took it a step further: everything we do is simply a behavioral reflex, an unstoppable response to external stimuli. Free will is an illusion. In the alleyways of a maze, a rat's performance is simply the end product of a chain of simple reflexes. On its first run through a maze, a rat is slow. It makes errors, lots of them. False starts. Dead ends. But a rat learns quickly. If you condition a rat to take a certain route through the maze by giving it a reward at the end of it, every time the rat runs through the maze it makes fewer and fewer wrong turns. The end result is a rat that runs quickly through a maze, turn after turn, directly to its goal.

In 1948, Tolman wrote: "The rat's central nervous system, according to this view, may be likened to a complicated telephone switchboard." In other words, the rat is a simple machine. Information about its environment is received by its sense organs, and outgoing messages are then relayed to its muscles. This was behaviorism. Ivan Pavlov—hard-eyed and white-bearded, like a grim-looking Santa Claus—had famously conditioned dogs to begin salivating at the sound of a bell. He was awarded the Nobel Prize for his work in 1904. He was a behaviorist. In the 1930s at Harvard University, prominent (non-bearded) behaviorist B. F. Skinner conditioned rats and pigeons to respond to specific cues in a specially designed box known as a Skinner box, or operant conditioning chamber. In the box, animals responded to stimuli by pushing levers or keys. Skinner thought free will was an illusion. We were machines responding to the environment. But Tolman disagreed. He wasn't a behaviorist. Instead, he called himself a field theorist.

With a series of maze experiments beginning in the 1930s, Tolman had made some surprising discoveries. He showed that behaviorism couldn't explain everything. Some things are true even if they can't be observed. For instance, Tolman trained rats to take a particular route through a complex maze to a food box. Then, he blocked the path to the food. Tolman discovered that a rat will take an alternative route through the maze—a detour around the obstruction. In other experiments, rats could find shortcuts through the maze, taking novel routes they hadn't taken before. Tolman proved that rats don't just learn a series of turns to get to their goal: left, left, right, left, food. That's behaviorism. Instead, the rat builds an informative spatial representation of the maze in its head. It builds a mental map. It was a field theorist's dream result. Behaviorism wouldn't have predicted it. A simple machine wouldn't have done it. "We believe that in the course of learning, something like a field map of the environment gets established in the rat's brain," Tolman concluded. "And it is this tentative map, indicating routes and paths and environmental relationships, which finally determines what responses, if any, the animal will finally release."

Tolman called it the *cognitive map*—the same mental map my wife uses to navigate through central Chicago, and someone else might use to find her way in an unfamiliar building.

It was a revelation. It turned behavioral psychology on its head. In

1948, Tolman published the results in a landmark paper titled "Cognitive Maps in Rats and Men." The discovery should sit alongside other great scientific discoveries of the twentieth century. It rivals splitting the atom, or the discovery of the helical structure of DNA. With his maze studies, Tolman made a free dive into the depths of the subconscious, where the light doesn't go. His work fundamentally changed how we think about one of the most basic aspects of our existence: how do we know where we are?

•

A cognitive map is different from a topological map, which is a much simpler representation of space. The positions of different points on a topological map are accurate to an extent, at least relative to one another—but the distances, directions, and scale on a topological map do not represent reality. It is the lie that tells the truth. The schematic map of the London Underground that was designed in 1931 is an example of a topological map. Intended to convey information to confused commuters, it is, for me, incredibly satisfying to look at: as rectangular and orderly as a circuit board, the map is crisscrossed with ranks of colorful straight lines, which are parallel, perpendicular, and diagonal to each other, and punctuated with stations. But it's not real. In a Euclidean sense, the map is nonsense.

On the map, the Central Line, as red as an English postbox, bisects the city like an arrow flying straight from east to west. In real life, the Central Line meanders across London. No part of it is really straight. On the map, even the River Thames—looping and serpentine in real life like a blue ribbon—has been tamed. The pre-1931 subway maps were geographically much more accurate, but impossible to comprehend. There are modern accurate maps of the London underground system online. They're overwhelming. Real London is a mess, a plateful of multicolored spaghetti. The London Underground map fixes this. It simplifies reality. It doesn't reflect the truth because that's not what topological maps are supposed to do.

A cognitive map of your environment is much more informative.

With sufficient information, it's Euclidean: it represents distance, scale, and direction as we experience them in the real world. We can consult it to calculate a precise route to a location in the same way that Tolman's rats used it to navigate a shortcut through a maze. Herein lies the power of the cognitive map.

Nora Newcombe is a cognitive psychologist at Temple University in Philadelphia. She studies how different people build their mental representations of the world. She has found that not all mental maps are equal. In fact, a third of Newcombe's subjects don't seem to form any kind of cognitive map at all. "Why do people think that there would be one answer that fits everyone, whether it's rats or humans or whatever?" says Newcombe (a 7, who says she has improved with practice from a 3 or 4). In Tolman's maze experiments, some of the rats performed badly, says Newcombe. The same is true for humans. Usually, we don't pay attention to these poor performers. Tolman wasn't interested in them. In experimental psychology, researchers often reduce the data of a group of subjects to an average score, or the mean.

"Most experimental psychologists see differences as an error," she says. Subjects are discarded if their scores are different enough from the group average. They're treated as outliers and removed. They are the noise that otherwise drowns out the signal. But Newcombe is interested in the outliers—in the variability around the mean. She wants to listen to the noise.

She tests subjects as they navigate Silcton, a virtual environment, using a method known as the *route-integration paradigm*. Seated at a computer, they navigate virtual tree-lined streets, gliding past blocky cartoonish parked cars. Gradually, they are taught two different routes through the environment. Along each route, subjects learn the locations of four distinctive buildings, or landmarks. Then, Newcombe shows her subjects two new pathways that connect the two routes they've already learned. She wants to know how difficult it is for the subjects to integrate the two environments—to merge them, to stitch them together to form a larger and more informative internal map. For instance, standing outside Building A on the first route, can a subject turn and point accurately to Building D on the second route? This would require them to access a complex internal representation of the virtual city—to imagine a pathway they have never taken. Not everyone can do it.

Newcombe's subjects fall into three roughly equal-sized categories, she says. One group represents people Newcombe calls the *integrators*. They navigate well, merging and combining the information from each environment to form a new super map. Another group does less well. They are the *non-integrators*. "The non-integrators do know the routes pretty well," says Newcombe "They just don't really fix the two of them in relation to each other." Then, there's another group: the *imprecise navigators*. The word *imprecise* is an understatement here. The data suggest they don't build an internal map at all. They're mapless. "Basically, the imprecise navigators don't really even learn the route," says Newcombe. "You can't really know the relation between the routes unless you know the route."

They're like me. They wander aimlessly around Silcton. They don't form a cognitive map. Newcombe says she doesn't know why they don't. "It doesn't really surprise me," she says. "In our particular kind of society, navigation skills are not that important—and not only because now we have GPS. Even a few decades back, if you knew where to catch the bus and, and maybe you had to change to a trolley or whatever, and then ended up at your place of work, you didn't really have to know where you were. You could develop those habits. When we were a bit more nomadic and had to look around for things to eat, it probably mattered more."

There are reasons why some people fail to construct a cognitive map, says Véronique Bohbot (an 8 out of 10). Not everyone uses the hippocampus—the diminutive curling seahorse, like a tissue-locked little comma in the midbrain—to navigate. A memory researcher at McGill University in Montreal, Bohbot has shown that some people use a brain structure known as the *caudate nucleus* instead.

And that's a problem.

Since the 1990s, neuroscientists have been using a powerful technique known as functional magnetic resonance imaging, or fMRI, to identify which brain regions perform different cognitive tasks. It can be used to map brain activity. On a neural level, activity is an electrical event: a neuron fires, which generates an impulse and allows it to communicate with other neurons by releasing neurotransmitters across the synapses that separate neurons from each other. When this happens, oxygen-rich blood flows locally to the active neuron. In the brain, blood flow equals activity. The changes in blood flow are something that can be detected and visualized.

On an fMRI scan, as the subject lies in an fMRI tube, an activated cluster of neurons shows up as a misshapen island of color in an otherwise black-and-white brain. The neuroimaging field has given these colorful regions of activity a decidedly nontechnical but descriptive name: they're blobs. A blob looks like a pixelated tangerine-colored splotch—the colored squares that form the blob are called *voxels*—that sits atop the monochromatic folds and ridges of the brain, like a direct hit from a paintball gun. There are limitations to the technique that have kept scientists squabbling for decades about how exactly to interpret the information collected from an fMRI scan. But, broadly speaking, this is what neuronal activity looks like.

In a study from 2003, Bohbot asked subjects in an fMRI scanner to navigate a virtual radial arm maze. In each arm of the maze there was a different virtual object. On the outskirts of the maze, subjects could make out the features of a simple landscape: the uneven peaks of a mountain range in the distance, or trees, or a setting sun.

After a training period, during which subjects explored the maze and learned its layout, Bohbot asked them to navigate to a certain object. *Go to the correct arm of the maze*, she told them. Around half of her subjects used the landmarks to navigate. They were guided by the mountains, and the tree line, and the sun—what Bohbot calls a *spatial* strategy. As they maneuvered through the maze, she saw an increase in their hippocampal activity. But the other subjects were different. They used *nonspatial* strategies, like counting the arms of the maze as they moved from their starting point. In these subjects, the hippocampus remained silent, but the caudate nucleus burst into sudden activity.

The differences between the two strategies are enormous. Among its other functions, the caudate nucleus is the brain's autopilot. Consider the following: you leave work and drive home along the route you've taken a thousand times before. When you get home, you realize you have no memory of the journey. That's because your caudate nucleus got you home. With repetition, the route has gradually become a procedural memory—like tying your shoelaces, or riding a bike. It's become automated. The brain delegates these kinds of tasks to the caudate nucleus to be efficient.

But it's not the same as navigating.

At the beginning of the study, around half of Bohbot's subjects used their hippocampus and navigated spatially. The other half employed a nonspatial strategy instead—the caudate nucleus. But, as subjects learned the layout of the maze, those numbers gradually began to change. With practice, more and more subjects stopped using a spatial strategy and put the burden on their caudate nucleus instead. This is how the brain *should* work. But, since the caudate nucleus doesn't build a cognitive map, it can be only just so informative. It won't reveal a shortcut through an environment. It can't suggest an alternative route home when the usual path is blocked by a storm-felled tree. Only the hippocampus can do that. And the hippocampus doesn't do it for everyone. Who are the people who navigated Bohbot's virtual maze using the caudate nucleus from the beginning?*

They're the mapless. They're the same people who entered Silcton, Newcombe's blocky virtual world, and wandered aimlessly through it, failing to make a map.

•

"Almost all of the work done on the cognitive map has been done in things navigating around space," says Tim Behrens, a computational

* Bohbot didn't test the subjects to see if they were flexible in the strategy they use to navigate. It's not known whether someone who uses her caudate nucleus to navigate could be trained to use her hippocampus instead.

neuroscientist at the University of Oxford. Mostly, scientists have been watching rats and mice in mazes, or moving along a track. For his work, Behrens (a 4 and "pretty rubbish," he says) tries instead to understand complex cognitive processes—like the formation of the cognitive map, or memory storage—by representing them as mathematical models. The cognitive map models more than just space, says Behrens. In fact, to even call it a map is an oversimplification. Instead, it's a sophisticated organizing machine; a model of the world; a record engine; an instruction manual; a pattern detector. It allows us to extract semantic knowledge about the world, says Behrens, so that we can understand how different parts of our environment relate to one another. "We call it structural knowledge," he says.

For instance, consider a movie like *Kill Bill*, or *Pulp Fiction*. "Or *Memento*, or something like that," says Behrens. They're all episodic movies, told from multiple different perspectives, with abrupt and jarring jump-cut transitions between scenes. They have nonlinear timelines. *Memento* is told backward. Despite this, we understand it. We can decode its narrative. In other words, says Behrens, from the abstract jumble we're given, we can extract specific information—the structural knowledge—rearranging the nonlinear narrative so that we can make sense of it. It's a subliminal process we take for granted, but it's powered by the same neural networks that form the cognitive map. "In those films, there are a lot of different events presented to you individually that are not presented in a way that's pieced together into a story," says Behrens. "The director allows you to do the piecing. But in order to do it, you need to know loads of background knowledge—like dead people don't eat breakfast, and that kind of stuff."

The same brain regions that form, and consolidate, and store memories—the cortex, and the hippocampus and its related structures—are responsible for providing this structural knowledge. "We learn the exact nature of how things relate to each other and the statistics of how things *tend* to relate to each other," says Behrens.

In other words, the cognitive map is our brain's director. Behrens continues: "I am currently sitting on my chair with my feet on the table, with a door to the left of me, and a sofa in front of me, and speaking on the phone to you. But, in my mind, that particular image means nothing to me, except when viewed through the lens of my previous experience where I understand that telephones are a way of communicating with distant people, and I understand what it is to be to the left of something, and I understand what it is to sit on something."

Computer scientists like Behrens can build powerful algorithms that model complex human abilities. For instance, modern AI computer programs can beat any human at chess, and Go, and *Space Invaders*. They can process and sort through millions of different move combinations in fractions of a second, and always make the correct choice. Deep convolutional neural networks have resulted in powerful—and slightly sinister—facial recognition software. But those computers still don't understand the world in the same way humans do. A computer can't draw the same rich, deep, intuitive inferences about the world that humans do.

"There's some magical way in which the brain or the cognitive map, or whatever, can abstract the key principles from some event," says Behrens. His computer models are complex and multilayered, but they fail to do this. The process requires an organic supercomputer like the human brain, endlessly gathering data about the environment. The hippocampus, via its connections to the cortex and other brain regions, then constructs a cognitive map allowing us to gather knowledge from the regularities in the world and make sense of it. But dark mysteries glimmer in the brain. "Somehow the brain is building a model of the world so that it can—inside its own cognitive space—figure out what the right thing is to do."

Behrens still doesn't know how the brain performs this task—this incredible leap of understanding. "It does it in this extraordinarily powerful way," he says, "and we have no idea how it works."

Chapter Three

In the Firing Fields

As a postdoctoral research fellow at University College London in the 1970s, John O'Keefe was interested in the hippocampus and its role in memory—like everyone else. Around that time, researchers had found a novel way to record the electrical activity of single neurons, by implanting a tiny recording electrode into the brain of a freely moving rat. When neurons are active, they generate a distinctive electrical signal—a spike known as an action potential—that can be measured if the electrode is near enough to detect it.

Working this way, O'Keefe believed he would gain important insights into memory. "I was going to go and see what memories looked like," he recalled, in a 2014 lecture at SUNY.

But that's not what happened at all. When O'Keefe positioned his recording electrode in the hippocampus and began to monitor the telltale spike patterns of neuronal activity, he detected two distinct populations of cells. One of them was predictable, firing in a regular and slowly rhythmic wave pattern, known as theta activity. But the second cell type was different. Most of the time, the second population of cells was conspicuously silent. They did nothing. But occasionally, one of them would burst into sudden activity, ramping up its firing rate into a noisy storm of electrical impulses—a steep mountain range of spike patterns. At first, O'Keefe didn't know why.

In 2014, he wrote: "[I]t was only on a particular day when we were recording from a very clear well isolated cell with a clear correlate that it dawned on me that these cells weren't particularly interested in what the animal was doing or why it was doing it but rather they were interested in where it was in the environment at the time." When the rat reached a certain location in the environment—for instance, the northwest corner of a large open enclosure—the cell fired: *click*. Elsewhere, it fell silent. When the rat returned to the location the cell had fired in before—*click*—it fired again. A cell that was active in the northwest corner of the box would fire in that location but nowhere else. As the animal explored its enclosure and O'Keefe watched the activity of the neurons, he realized: "The cells were coding for the animal's location!"

O'Keefe named them *place cells*.

•

Found almost exclusively in the hippocampus, place cells are a type of neuron known as a pyramidal cell, first described more than a century ago by the Spanish neuroscientist Santiago Ramón y Cajal. During his long career, Cajal rendered hundreds of finely detailed neuroanatomical images of different brain structures, showing their microscopic structure in exquisite detail. He was awarded the Nobel Prize in 1906 for his work. He made several important discoveries and brought the architecture of the brain to the page.

One of Cajal's intricate ink and pencil drawings from 1896 shows pyramidal cells from a rabbit cerebral cortex. They look like uprooted trees from a strange grey forest, their root structures floating above ground. A long, straight axon extends from a pyramid-shaped cell body before branching and bifurcating into a thick arbor of dendrites at each end, sharing local connections with thousands of other neurons both that inform it and that it informs. Pyramidal cells are found widely in the cerebral cortex, and in the amygdala, but they only seem to encode spatial location in the hippocampus, or nearby. To complicate matters, a few years after the initial discovery of place cells, O'Keefe described *misplace* cells. If an animal travels to a location in its environment expecting to find something that is absent instead, the misplace cell begins to fire.

O'Keefe showed that when the rat is at rest, a place cell fires once every ten seconds or so. But, when activated, it begins to signal much more quickly, a flurry of action potentials arriving at a rate of around twenty times a second or faster. These impulses act like a locating beacon, a cursor, a pin in a map. The precise location that a place cell fires is known as its place field or firing field. Imagine, for instance, that you are standing at your front door: a place cell activates. But as you step into your house and begin to walk down your hallway, that particular place cell stops firing. It quietens. It belongs to that single place only—to the front door. As you begin to move through your house, a procession of other place cells begin to fire in turn, one after another, from room to room, before falling silent again. The activity of each cell indicates a distinct location in your house. Cell #008: the kitchen sink; Cell #192: your favorite reading chair; Cell #417: the window in your bedroom that overlooks the street. And so on. In this way, place cells are endlessly mapping your entire spatial environment one location at a time.

But how do they do it?

•

"In the most straightforward sense," says Lynn Nadel, who coauthored *The Hippocampus as a Cognitive Map* with O'Keefe in 1978, "a place cell is a neuron typically in the hippocampus, although things like them are found elsewhere, whose activity is somehow modulated by, or caused by, or related to, where the animal is located in its environment." But that's not all it does, he says. In the same way that the definition of a cognitive map is being carefully revised, researchers have begun to ask whether place cells might have a broader role too. "Is it really what we think it is when we call it a place cell?" asks Nadel. "It actually may be something a good deal more interesting. People are beginning to talk about them not as place cells but as engram cells,* or concept cells." The debate about precisely how to define and think about place cells is likely to continue until neuroscientists reach a consensus—and perhaps they never will. For his part, Nadel thinks place cells are one component of a larger neural network. "They don't sit there all by themselves holding up a flag telling the animal: you're here," he says. "They're part of a broader network of cells that's really dealing with the sequences of actions that the animal is taking, and where those lead the animal, and what to expect when you get there."

When O'Keefe and Nadel published *The Hippocampus as a Cognitive Map*, it was a neuroscientific, philosophical, and technical manifesto. It was a game changer. Somehow, it was both lyrical and erudite. With it, an entire field of neuroscience was born. It began: "Space plays a role in all our behaviour. We live in it, move through it, explore it, defend it. We find it easy enough to point to bits of it: the room, the mantle of the heavens, the gap between two fingers, the place left behind when the piano finally gets moved."

From that simple and whimsical beginning, they then made a leap, asking a series of questions that, like Buddhist koans, leave my brain tied in a knot: Can objects exist without space? Can space exist without objects? If the space between two objects is actually filled with tiny particles, is it still space? Does space even exist, or is it an invention, a human construct—a figment of our imaginations? If we invented space, how did we do it?

These were the mind-bending and existential questions that began the search for place cells.

In 2014, O'Keefe was awarded a Nobel Prize for his work on the complex neural circuitry that controls navigation. He shared it with two Norwegian researchers for their later work on other cells that encode space. Now white-haired and past eighty, with his chinstrap beard

* An engram is a theorized change that occurs in neural tissue, which leads to the persistence of a memory—in other words, it's a memory trace.

intact, O'Keefe is still at it, working in the same lab fifty years later at University College London. O'Keefe and Nadel had graduated together from McGill University in Montreal in the late 1960s: "An Irish kid from the Bronx and a Jewish kid from Queens," as Nadel put it in a 2014 interview. Now they were in London together, working on the internal navigation system. Nadel had left his postdoctoral fellowship in Prague in August 1968, when Soviet tanks rolled through the cobblestone streets of the medieval city. Loading his then-wife and two kids into a van, he drove to O'Keefe, already in swinging London. They were the upstart Americans.

"We weren't looking for this particular form of activity," Nadel tells me. "When you first stick electrodes in the brain of an animal and you record under conditions no one's ever recorded before, you don't know what the hell you're going to see."

In the lab, O'Keefe and Nadel had rigged their recording apparatus to produce a sound every time a place cell near the electrode began to fire. Back then, data were recorded onto magnetic tapes and analyzed later. The location-specific firing patterns had taken them by surprise.

"The first time we heard it," says Nadel, "it was like: *What the hell was that?*"

•

When I call André Fenton on his cell phone, he has just stepped from a mid-morning train into the cool, cavernous, high-ceilinged bustle of Union Station in Washington, DC. The noise of other commuters is a steady tidal rush around him. A neurobiologist at New York University's Center for Neural Science, Fenton (7 out of 10) studies the storage and coordination of memory in the human brain. "I happen to be very interested in knowledge," he says into a wall of white noise, "where it comes from, how we get it, how we make it, whether it corresponds to things that are actually real, and so on."

Since place cells store a particular kind of knowledge—spatial knowledge—Fenton is interested in them too, along with the neural systems they help form. "The cool thing about the navigation system," he says, "is that it's a whole system of knowledge that we all get and we all use. We can prove that we have it by using it. I just got off the train at Union Station in Washington, and it wasn't random that I got here."

But to Fenton and many others, place cells still represent an unsolved riddle. "By where they discharge action potentials they seem to signal locations in space," he says. "Now, what's interesting in particular about what I just said is if you take another step back and say, 'Well, how would they know where their location in space is, to signal it?' "

It might be tempting to think that place cells are like the cells that make up other sensory organs, like our eyes and ears. But they're not. They're different in important ways. Consider the eye: the retina at the back of the eyeball acts as a sensor for light. Visual information is gathered when light falls on the specialized cells there and is transmitted via neural pathways to the brain, where we can begin to make sense of it. The visual cortex then orders the sensory information gathered by our eyes. It edits and interprets that information for us. Sight is complicated enough, but at least it begins with input from the physical world: light.

Light is tangible. "You can trace it to the real world, at least in principle," says Fenton. "The cool thing about place cells is: you can't. We explicitly don't have a sensor for locations in space, yet these cells seem to know something about locations in space." Place cells remain a mystery. Fifty years since they were named, we still don't fully understand them. Almost all of what we do know has come from animals in a box, or a maze, or running along a track. Place cells are supple navigators. They allow us to map any location on the planet. They are powerful beyond measure. When humans finally travel to Mars, says Fenton, our place cells will allow us to navigate there too. They map the entire universe. They even allow us to explore imaginary and virtual places—locations that don't exist at all. "You probably understand Hogwarts," says Fenton, "and it doesn't exist." In rats, place cells continue to build a cognitive map even when the animal is in darkness. The place cells even fire in a location-specific way if a rat is fitted with a miniature blindfold—a fact that is as ridiculous as it is informative.

How can place cells do this? Fenton says there are relatively few of them. How can they compute and encode an infinitely large universe, and even encode location for nonexistent and imagined places? In fact, Fenton explains, it takes more than a single place cell to signal a location. Many more. A rat exploring a small open enclosure might need only a handful of place cells to encode its location, but in a larger and more complex environment, more place cells are needed. This is where the numbers are important.

Fenton says: "One way of thinking about this is, there are, let's say, in the order of a million cells in your, or a mouse's or a rat's brain in the hippocampal system, and there are different parts of that system." In each part of the system, says Fenton, there are a couple of hundred thousand place cells, and approximately ten percent of them are active at any moment in time. As an individual moves around an environment, a different ten percent of place cells becomes active, firing to represent a specific location in space. "They don't become active in a simple way, like on a checkerboard—first this set, and then a completely different set one step over," says Fenton. "It's a continuous representation. There are ten-thousand-ish place cells firing at any one moment. At every place in the universe a unique ten thousand cells will be firing."

In other words, the place cell that fires, bursting into activity when I stand at my kitchen sink—Cell #008—is unique. But it has an estimated 9,999 or so comrades simultaneously firing with it, scattered throughout the hippocampal system and possibly beyond its borders too. When I sit in my favorite reading chair, another 10,000 place cells fire—a totally different combination of cells that encode my position. Perhaps some of my place cells fire in both locations. But others don't.*

It's the specific combination of place cells firing in concert that represents a place. This organizing principle is called an *ensemble code*, since it requires a discrete and unique ensemble of place cells firing together at once in an orchestrated event—a synchronized burst—to encode a single location. The computing power of a system like this is

* There is some evidence that the ensembles of place cells representing a specific place might drift over time—a concept known as representational drift.

incredible. And bewildering. If there's a pattern to the way that place cells fire together—to what determines a specific ensemble—scientists haven't found it yet. There's no topographical relationship between two place cells. In other words, two place cells that sit next to each other in the hippocampus are as likely to represent two distant locations in an environment as they are two locations that are near to each other. They might both fire in the same location, as part of an ensemble. Or they might not.

"Just like you can compute, with an alphabet of twenty-six letters, a very, very large number of words," says Fenton, "you can compute, with a small number of these cells, or a relatively small number—a few hundred thousand—virtually an infinite number of possibilities of location."

Computational neuroscientists have a name for the principle by which a relatively small population of cells—for instance a few hundred thousand place cells in the hippocampus—fire together to encode something vast and infinite, like the physical universe. It's known as *sparse coding*.

If Fenton wants to learn something about place cells and how they encode our position in space, he must first insert a recording electrode into a brain to monitor the electrical activity of place cells. It's the same technique that O'Keefe was using in 1970. Usually, researchers use rats or mice for this work. Almost exclusively, they aim the electrode at the rat's hippocampus, the brain region where place cells are particularly abundant. This isn't an easy thing to do. Gradually, though, over the past few decades, neuroscientists have become very good at it.

For more than a decade, researchers have been using use tetrodes, each of which has four separate electrodes on it. This way, they can record the firing activity of several different neurons at a time, the way a microphone dropped into a cluster of people can record several threads of conversation at the same time instead of just a single voice. Even so, because place cells are scattered throughout the hippocampus, Fenton can only monitor a few of them at the same time—perhaps as few as ten in one animal, he says. If he's lucky, his electrodes might sit close enough to as many as sixty place cells at once. He can watch them firing together in real time as the rat moves around. But since there are a few hundred thousand place cells in the hippocampus, and a few scattered beyond its borders too, if it takes the sudden synchronized firing of an ensemble of around 10,000 of them to encode a specific location, as Fenton suspects, even the best study provides an incomplete picture. It's a little like studying the dynamics of a rioting crowd by tracing the movements of a handful of people in it. Or piecing together a conversation between 10,000 people by listening to just fifty voices.

•

In some very important and fundamental ways, place cells remain a mystery. Researchers haven't even reached a consensus on how many place cells a normal brain contains. Fenton, like everyone else, is working with the best estimates that research has provided. Even so, researchers can construct a physical map of a rat's environment based on the firing of the individual place cells: another kind of cognitive map. Essentially, the place cells *are* the map. By determining where in a specific environment a handful of place cells fire, researchers can map out the firing fields of each cell. For instance, Place Cell #7 only fires when the rat is exploring the northwest corner of the enclosure; Place Cell #13 is only active when the animal strays into the center of the box. Once a scientist like Fenton has constructed the rat's cognitive map—determining the firing fields for specific place cells—he no longer needs to see the rat to know where it is in space. Instead, he can reconstruct its physical location from the signaling of its place cells—a process known as decoding. He's reading the rat's cognitive map directly from the cells in the rat's brain. He's reading the rat's mind.

There are videos on YouTube that show the process in action. In one grainy black-and-white video, captured by Matt Wilson's lab at MIT, a rat runs along a narrow, winding, elevated track. The camera records the rat's movements from above. As it runs, researchers are recording the activity of seven different hippocampal place cells at once. Wilson has color-coded the activity of each of the seven neurons to be able to distinguish the activity of each of them. At the beginning of the track, one of the place cells is firing much more rapidly than the others, each impulse accompanied by a crunchy little clicking sound. This neuron is represented by a green dot superimposed onto the track on the exact location it fired. This neuron's activity is isolated to the beginning of the track. In other words, of the seven neurons Wilson is monitoring, this one—we'll call it Place Cell #1—only fires at this location. As the rat takes a wide right turn on the track, the green neuron falls silent and another place cell—Place Cell #2—begins to fire instead. A blue dot has been superimposed onto the track to show the rat's position when the cell fired. A string of blue dots appears as the rat completes the turn in the track—in this way, we know its firing location. But it's far easier to list what we don't know about place cells than what we do, says Fenton.

"I think we don't understand them at all," he says, laughing.

But, just like the cognitive map they help to inform and construct, place cells encode more than just space. Instead, Fenton believes they store knowledge—all sorts of knowledge. Recent research shows that cognitive maps even allow us to navigate abstractions like social space. In the hippocampus, in addition to mapping space, we construct complex maps to understand social hierarchies and help us navigate social dynamics. In fact, there's nothing particular about a place cell that even distinguishes it from other hippocampal neurons that aren't place cells. Perhaps all hippocampal neurons represent spatial information under the right circumstances. Other studies have shown that place cells encode other nonspatial information, like sound, or odor. Or faces. Or objects. We think of them as place cells simply because almost all the information we expect them to encode has a spatial component. "I imagine that they are knowledge representers, or something," says Fenton. "And the knowledge is totally abstract. These are things that are computed in the mind."

This explains how we can use them to map imagined places, like Hogwarts, or real but uncharted places, like Mars. Despite their computing power, place cells are fallible. They misfire, says Fenton. If we depended on a single place cell, or even just a few, to tell us where we were, those misfirings would be easier to detect. Requiring 10,000 cells to fire at the same time to represent a place is a measure against this. I ask Fenton a question I have been turning over in my mind: how is it possible that place cells never seem to encode the wrong location?

They do, says Fenton. It happens all the time. You could be standing in a rock-strewn Scottish bay, leaning into the endless rain, and your place cells might instead signal that you're in Times Square, or Tokyo, or standing in your shower at home. In these moments, he says, they're encoding recollected places. Memories. Previously visited locations. "This is of course what one would want of a useful place-representing system," says Fenton. "How could you ever know about *there* if there was never represented? If you can't know about there, you cannot plan or navigate with purpose."

Fenton can record these instances in real-time with electrodes. "If you look at enough place cells together you see that they are firing in the wrong place, as if pointing to or representing some other place," he says. "You can consider that noise, but if you do the right kind of experiment you can see that they're simply representing a memory of a place, or a place one is considering visiting, or not visiting, and so on." But place cells don't misrepresent location across minutes. They do it for half a second, or less. "It's happening all the time," says Fenton.

And if you look carefully, you can even see it happening.

•

A researcher at the Center for Integrative Neuroscience in San Francisco, Loren Frank is most interested in these moments—the fractions of a second when place cells are representing some other place. In those strange fleeting half seconds, says Frank (7 out of 10), ensembles of place cells are sorting through the countless possible locations we *could* be in too, and considering the almost-infinite array of different routes we might take to get to them. Like everything else in the brain, the signature for this sort of activity is electrical—and we can observe it, he says. It's known as the sharp-wave ripple. Neuroscientists first recorded sharp-wave ripples in the brain as early as 1969 with the first primitive recording electrodes. John O'Keefe saw them in 1978, while monitoring cellular activity in his freely navigating rats. Decades later, researchers have now observed the same distinctive patterns of electrical activity in all mammal species they've looked for them in so far: in humans and other primates, and cats, and bats, and rodents. Sharp-wave ripples have even been observed in nonmammal species like zebrafish. But what are they?

"They tend to happen during what you might think of as more offline periods that people might associate with just mind wandering, or relaxed thinking perhaps," says Frank. Detected by a recording electrode, a sharp-wave ripple looks like a single large electrical spike followed by a series of rapidly diminishing oscillations—like the ripples on the surface of a lake after a stone has been thrown in. They only last for a hundred milliseconds or so, but they're unusual because they can involve up to 100,000 neurons at once, all firing together in one of the most synchronous firing patterns in the brain. "Historically, they were actually studied more during sleep," says Frank, "with the idea that the hippocampus was taking patterns it had experienced during waking and then replaying them over and over again to drive plasticity outside the hippocampus for long-term memory storage." For this reason, neuroscientists have associated sharp-wave ripples broadly with the half-understood process of long-term memory consolidation, and sometimes refer to them as replay ripples. Predictably enough, studies have shown that disrupting sharp-wave ripples has a profound negative effect on long-term memory consolidation.

Neuroscientist György Buzsáki (8 out of 10) was the first person to systematically describe sharp-wave ripples in a 1983 paper. "You have about 3,000 sharp-waves in a night," says Buzsáki, now a researcher at New York University Langone Health Neuroscience Institute. "If I erase all 3,000 sharp-waves from your brain, leaving everything the same as it was before, then you wouldn't remember this conversation."

But sharp-wave ripples are involved in more than just memory. Sometimes, says Frank, entire ensembles of place cells—thousands of neurons at once—fire together in synchronous bursts across the hippocampus.

"What was really exciting to us about sharp-wave ripples, and still is exciting to us," says Frank, "is that they express often coherent sequences of spiking activity that correspond directly to a path, or a possible future experience." In other words, sometimes an ensemble of place cells fires together to signal that an animal is somewhere it is not, just as Fenton has seen. In these moments, place cells are not malfunctioning. Instead, they're reexamining spatial memories from the recent past, and sorting through a multitude of different possible futures, like the spaghetti model that shows the possible routes a hurricane might take. At any moment, many spatial futures exist.

This makes sense to Buzsáki, the sharp-wave ripple expert. He has suppressed sharp-wave ripples in sleeping mice after they have explored a maze. Normally, during sleep the mice would be busily consolidating their spatial memories, building an accurate cognitive map of the layout of the maze. But not without ripples. Buzsáki saw that interrupting hippocampal ripples was as detrimental to memory as removing the entire hippocampus. Without them, the mice are lost.

"The hippocampus is not only involved in navigation," says Buzsáki. More broadly, as we know, it's the memory powerhouse of the brain. But it also performs a more complex kind of navigation—Buzsáki calls it mental travel. "Mental travel can go back to the past," he says. "We call it memory. It can go to the future, and we call it imagination or planning."

Back in his lab in San Francisco, Loren Frank has been studying these brief instances of mental travel—"that is," he says, "when the place cells, given their general activity, are active in a way that suggests the animal isn't coding for where it is right now. We published a paper that showed the animal could be awake in one environment and replay a sequence that was uniquely associated with another environment it had been in, say, twenty minutes ago."

In fact, when you start to look for it, he says, this kind of activity in the brain is not even that uncommon. "It turns out there's a surprisingly large fraction of the time when the system doesn't seem to be representing current position," Frank says, "as though it's thinking about other places."

When I spoke with André Fenton weeks earlier, he'd mentioned these moments too, as he rushed across the platform at Union Station. I had imagined him walking beneath the vaulted ceiling and across the checkerboard tiled floor of the station—a place I have been before. For a moment, the unique ensemble of place cells in my brain that encodes Union Station fired together in a burst of spatial memory. In other words, I mentally traveled there.

In 2020, in the journal *Cell*, Frank published an article with the *Matrix*-like title: "Constant Sub-Second Cycling between Representations of Possible Futures in the Hippocampus." In it, he and his coauthors explain that as we navigate, our place cells switch rapidly between encoding our present location and representing where we might go next—different ensembles of thousands of cells alternating between the *now* and an array of possible futures, shifting constantly as we hurtle toward them. Back and forth. From the present to the future. At a rate of roughly eight times a second, as a rat navigates a three-arm maze, its place cells are considering the rat's possible future positions in space. "What the animal can do, as it's running up toward a decision point, is play out trajectories that go either to the left option or the right option and it can alternate between them eight times a second," says Frank. "So it can go: up to the left, up to the right, up to the left, up to the right, and then it goes one way or the other."

The brain propels its owner into possible futures, surging forward into imagined or remembered spaces. Place cells perform this mental travel around twenty times faster than physical travel occurs in the real world. "That's incredibly important," says Frank, "because if your memory can't work faster than your body then it's not very useful. You want to be able to think about a place, and you want to be able to just zoom out to it." From an evolutionary point of view, it's important to be able to predict what might happen in the immediate future because it allows us to make good decisions. It keeps us safe. It allows us to plan ahead. Earlier humans might have wondered if a predator was waiting around the next bend in the path. The modern equivalent, says Frank, is a question like "What route should I take to work today?" Sharp-wave ripples allow us to mentally simulate the future possibilities and model the best routes.

"If you have the ability to use your past experience to imagine the outcomes, that is a huge advantage in terms of making better decisions about what to do next," he says. "The reason evolution created systems that can store memories is not because reminiscing is so useful, but because those memories are incredibly useful for making more accurate predictions about what will happen next."

When the hippocampus glimpses the possible futures, it enlists the whole brain to make sense of them, to assess and consider them. "In the prefrontal cortex"—the brain region that governs executive functions, such as decision making—"we see up to maybe seventy percent of the neurons that we randomly select changing what they're doing at the time of these sharp-wave ripple events," says Frank. He sees the same changes in the nucleus accumbens, a part of the brain that is important for associating experiences with their outcomes, with about half of the neurons he randomly selects undergoing a sudden change in activity levels.

"It's a brain-wide phenomenon," he says. "We want to understand this idea that the brain doesn't always think about the now."

In a complex environment—navigating the steep alleyways of Montmartre in Paris, or a difficult section of a mirror maze in Chicago—place cells are always weighing different options, following future trajectories: Left at the fork, right at the fork, left at the fork, right at the fork. Do we think the human brain works in the same way?

"Absolutely," says Frank. "That's how we think it works."

He suspects that these brain events could even help to explain why some people are better navigators than others. Early results from his ongoing research suggest that not all people generate sharp-wave ripples in the same way. In other words, some people are better at projecting themselves into their future spatial possibilities—at surging ahead into mental space. Perhaps my wife—by generating more, or perhaps stronger, sharp-wave ripples—is simply better at weighing the possible routes through an environment and predicting the best one to take.

"The speculation could go something like this," says György Buzsáki, when I ask him the same question later. "There is a continuum of everything; there are no discrete things in the brain. When it comes to creativity, it's a continuum. When it comes to perceptual skills, it's a continuum. It's common knowledge that some people are better navigators than others. Whether we come with preconfigured brains to be good or bad . . . that's the $64,000 question."

•

There's a reason most of the data on place cells have been collected in rats and mice and other animal models. Put simply: humans won't volunteer to explore an environment while a surgically implanted electrode records the changes in their brain activity. If they did, it wouldn't be permitted. Even so, there are moments when it's possible to place a recording electrode into a human brain, says Arne Ekstrom (6 out of 10), a cognitive neuroscientist at the University of Arizona. For decades, he's been recording the activity of place cells in epilepsy patients.

"I've probably tested, in my career, a hundred or so patients who have had electrodes in their brain," says Ekstrom. "Electrodes are placed in many different locations in their brain to try to better determine where the seizures are coming from." While the patients are in the hospital, essentially waiting for another seizure to arrive and for the electrodes to determine its source, Ekstrom asks them to take part in virtual reality navigation tasks. He seats them at a laptop and synchronizes the computer program and the data being collected by the recording electrodes embedded in the patient's brain.

"That way," he says, "we can say that when you navigated this area in virtual reality on the laptop, we know it related to action potentials or some kind of other electrical activity that happened when the hospital was recording the activity."

Ekstrom's work has shown us that humans have place cells too. He says around twenty percent of the hippocampal cells he monitored seemed to be place cells, firing based on the subject's location in space. "I've never put too much stock in that number," he says, "because I think there could be variability depending on how demanding the task is. And there could be variability based on the electrode placement."

•

With fMRI technology, neuroscientists have been able to map the brain and its multitude of functions. We now know the precise location of the fusiform face area, a nondescript little knot of cortical neurons located on the underside of the brain that responds with a flurry of neural activity when a person entombed in a scanner is shown an image of a human face. These cells are just hardwired to respond to faces. Subjects with a relatively thicker cortex at the fusiform face area than other people are better at recognizing faces.

Other studies have looked at aggression, and depression, and dreaming, and drug addiction, and grief, and olfaction with fMRI. For a study titled "Tracing Toothache Intensity in the Brain" researchers placed subjects in an fMRI tube and then hurt them with electric shocks to the teeth. A 2009 *Journal of Cognitive Neuroscience* article investigated the neuroanatomy of anger and angry rumination. The paper matter-of-factly explains in detail how "participants were insulted and induced to ruminate." Once in the tube, unsuspecting subjects were given anagrams to solve, but then, the researchers wrote, "As part of the provocation manipulation, the experimenter interrupted participants thrice requesting that they speak louder. During the third interruption, which served as the anger induction, the experimenter stated in a rude, upset, and condescending tone of voice: 'Look, this is the third time I have had to say this! Can't you follow directions?' "

Alone in the tube, the subject is left to ruminate over an unwarranted insult while the machinery of the scanner steadily hums away, collecting the forming blobs and voxels of unresolved fury. As a result of this study, we know that anger resides deep in the dorsal anterior cingulate cortex, but angry rumination burns away in another part of the brain, glowing like a hot coal in the medial prefrontal cortex. The blobs of romantic love are scattered throughout the ventral tegmental area and other parts of the brain's reward network. But the effect of lovelorn feelings—the dull insistent sadness that follows a romantic breakup—can be detected by a reduced activity in the caudate nucleus, like the dimming of a light.

University College London researcher Eleanor Maguire (who awards her spatial skills a 1) has spent decades using fMRI and other neuroimaging techniques to determine the brain regions involved in navigation. Her experiments go something like this: a subject lies in the scanner as researchers provide a spatial task to perform—study this map, for instance, or mentally calculate the best route between your home and workplace. When a subject navigates, even lying horizontal and imprisoned in a scanner, specific brain regions light up like a fireworks display. During navigation, the hippocampal formation is blob central. In an early imaging study from 1998, titled "Knowing Where and Getting There," Maguire investigated navigation with positron emission tomography, or PET—another approach that provides information both about the structure *and* the function of an organ.

Not only is the right hippocampus incredibly active during navigation, Maguire showed that the more accurately a test subject navigates, the more active it is—the larger and brighter the blobs. In another part of the study, subjects were asked to perform another navigation task, in which they navigated toward an unseen goal. This time, a different part of the brain lit up—the right inferior parietal cortex. These two parts of the brain are doing two different parts of the job of navigation. Maguire says the hippocampus provides the representation of space: the cognitive map. But the inferior parietal cortex is computing the route—mentally navigating the cognitive map—and calculating the sequence of turns a person must take in order to reach the goal. The left hippocampus is active too, but at a constant and unchanging level. Maguire believes it maintains an unchanging memory trace of the route. In another trial, subjects in a scanner navigated a virtual environment and found the route blocked. This time, parts of the frontal cortex—the brain region involved in planning and decision making—light up instead.

Among humans, some of the best navigators in the world are London taxicab drivers. They have to negotiate the complicated vascular system of an enormous sprawling modern city. Look at a map of London again: it's terrifying. The London Orbital Motorway, or M25, encircles the city. Within the 117-mile-long ring road, the landscape is riddled with more than 25,000 roadways. Streets radiate everywhere, intertwined, labyrinthine—almost 10,000 miles of road. A London cabbie must know them all. They are tested on their knowledge, sitting for an arcane exam—literally known as The Knowledge—before they can obtain the license to helm a London taxicab. Even since the advent of smartphones, London cabbies must absorb The Knowledge.

A typical candidate studies for as long as four years to memorize everything completely: the shortcuts, the one-way streets, the ring roads and bypasses, the bends of the river and the streets that follow them, the daily patterns of congestion, the tourist spots, the points of interest. During that time he will rack up around 50,000 road miles, zipping through the city on a motorbike with a worn map taped to the windshield for reference.*

Eventually, a cabbie holds the entire city in his head. In 2006, Maguire neuroimaged the brains of London cab drivers and compared them with those of bus drivers who drive the same streets. Unlike taxi drivers, bus drivers are trained over a period of just six weeks and they never deviate from relatively short, well-defined routes. Driving a bus doesn't require four years of independent study. Bus drivers don't need to store a sprawling city in their heads. They need to remember the endlessly repeated loop their bus makes. The brain differences were dramatic. The volume of a London cab driver's posterior hippocampus—the region most involved in encoding spatial information—is significantly larger than a bus driver's, and it's packed with more grey matter. Like a bodybuilder's bulging biceps, the hippocampus has responded to the constant training, to the stored spatial data—to the years of cutting through the city streets on a motorbike in the rain. The hippocampus has swollen in size,

* According to Transport for London's 2017 statistics, only 2.3 percent of London cab drivers are women.

adapting to the new expectations placed on it by the cab driver. The longer a cab driver has piloted his cab through the congested web of London, the more voluminous his hippocampus has become.

When I tell Emeline about the study, she tells me, *I'm a taxi driver; you're a bus driver.*

•

One particularly modern approach has become as important to researchers studying navigation as mazes were to Edward Tolman. It's what allows neuroscientists to study bus drivers and cabbies as they navigate London's streets; and it helps Arne Ekstrom to glimpse the neural workings of an epileptic patient who can't even leave her hospital room. It's virtual reality (VR) technology.

When Maguire studies taxi drivers, they're piloting a virtual cab through a virtual London. For an early VR study in 2001, University College London computational neuroscientist Neil Burgess (between 7 and 8) adapted the popular commercial first-person shooter video game *Duke Nukem 3D* for spatial research. In the original version—tagline: *Prepare Yourself for Total Meltdown!*—a heavily armed protagonist called Duke Nukem navigates a crumbling cityscape, defending it against an alien invasion. Burgess rewrote the code. In the modified version, study participants explore an empty—and mildly unsettling—environment measuring seventy meters by seventy meters. The alien overlords have left the city. Perhaps they won. Subjects recruited to the study navigate the unpopulated streets, gliding along a desolate main street, past an abandoned cinema and an empty bookshop. Like the amnesic patient H.M., thirty subjects had undergone selective removal of the temporal lobe as a treatment for intractable epilepsy. But, unlike H.M., only the left or right temporal lobe had been surgically removed—not both. Along with sixteen healthy controls, the subjects explored their virtual environment, walking the streets. Afterward, Burgess tested them on their navigation skills, and their ability to draw maps, recognize scenes, and remember the details of specific events that occurred in the VR city.

As predicted by Maguire's earlier work, people missing their right temporal lobe were much more impaired in navigation tasks than subjects whose left lobe was removed. When subjects were asked to depend on their cognitive map to navigate an environment, rather than simply follow arrows on the ground that pointed the way, the right hippocampus selectively lit up on scans, showing greater levels of blood flow in that brain region. In other words, there's an asymmetry in the brain. We're lopsided—with the right-hand side of the hippocampus doing the heavy lifting during navigation. The left side of the hippocampus is more concerned with episodic memory—with other important details, like the sequence of events that occurred during the navigation task, and the order they came in.

Even more amazing: the accuracy of navigation is proportional to the increase in blood flow to brain regions involved in wayfinding. If a subject took a winding and circuitous route to the target—if she wandered—her right temporal lobe was less activated than if she took a direct route.

Current virtual environments look very different from the empty city Burgess created in 2001. His city was simple and pixelated. With today's technology, subjects navigate virtual cities that seem completely real: the virtual wind flattens virtual grass; virtual water is punctuated by virtual ripples and reflects the towering virtual clouds that float overhead. "We work with immersive environments by having people navigate on an omni-directional treadmill, so we can basically produce whatever environment we want," says Ekstrom. "We can put you in the Sahara Desert. We can put you in the tundra."

It might seem unlikely, but researchers even use VR for animal subjects. There is a modified setup for mice. As an electrode records ongoing place cell activity in the hippocampus, a mouse runs atop a large Styrofoam ball called a JetBall, which floats and bobs on a cushion of air. Displayed on a wide screen in front of the mouse, a simulated environment—the striped walls of a maze, for instance—scrolls endlessly past. There is something oddly surreal and dystopian about watching a mouse gallop on top of a spinning globe in a darkened room, bathed in the unnatural light of an unfurling artificial world. But it's incredibly informative. As the mouse runs in place on the JetBall, the software gathers data on the activity of individual place cells at different locations in the maze. From this and other work, researchers are beginning to understand more precisely how place cells encode locations.

"I suppose when John [O'Keefe] discovered them it was mysterious how they knew how to represent space—what exactly was driving them," says Burgess, who was junior researcher in O'Keefe's lab in the 1990s and still occasionally collaborates with him. "Since then, I guess we've found out a lot more about the kinds of environmental inputs that make them fire in a given place."

In recent years, Burgess and others have begun to understand that place cells encode space based on a bombardment of environmental information, collected elsewhere in the brain. Most importantly, the spatial information that place cells generate must make sense, says Burgess. "It's wired up to produce a coherent answer," he says. "The most common idea is that all those place cells are connected together, so all the cells that should be firing in the same place support each other and inhibit the ones that shouldn't be firing, and so you get a lot of noise reduction or error reduction because all those, say, 100,000 neurons are working together to come up with a coherent representation of one place." Burgess adds, "The place with the most votes is likely to be the one that gets represented. It's not that they're all independently telling you where you are, which obviously would be susceptible to quite a lot of noise."

In other words, place cells are discerning and selective when they signal spatial information. It is their greatest gift: Coherence. Noise reduction. Imagine you're standing in an elevator: when the doors open, you see the concourse of a busy airport. People rush past pulling suitcases on wheels. This scene has logical consistency, or coherence. It makes sense. Elevator doors *do* open to reveal airport scenes. Next, imagine instead that when the elevator doors open, you see a remote and windswept beach. There's no logical consistency to that event. Even if two locations are represented by ensembles of place cells with lots of *indi-*

vidual cells in common, they will represent the location with the most logical consistency. Place cells vote for meaning.

•

Several decades ago, University of California, Santa Barbara cognitive psychologist Mary Hegarty (6 out of 10) developed a test she calls the Santa Barbara Sense of Direction Scale. By answering a series of questions, test-takers rate their own sense of direction. The questions include:

I have a good memory of where I left things.

I very easily get lost in a new environment.

I remember routes very well while riding as a passenger in a car.

Data collected from human subjects show the existence of a wide spectrum of spatial abilities. The results of Hegarty's studies show that, on a deeply intuitive level, we know if we're bad navigators. Some people navigate with confidence and skill while others—like me—struggle constantly. The only way to investigate the range of abilities is to test millions of different people, the entire spectrum. For researchers, this has remained impossible—or impractical, at least.

But Hugo Spiers has found a way to do it. A cognitive neuroscientist at University College London, Spiers (a 6) is interested in how the brain represents space, and the ways we navigate it. And he's using a modern approach: a cell phone app. His subjects play a game called *Sea Hero Quest*. It's the kind of game you might see someone playing absentmindedly in an airport, or waiting in line. Players try to memorize a map before piloting a boat called the *Sea Hero* through progressively more difficult virtual seascapes. Level by level, players trace circuitous paths to distant markers, on the way collecting the missing pages of a journal and photographing sea monsters.

I played *Sea Hero Quest*. And completely failed. Within minutes, I was tracing wide figure-eights in the water, searching hopelessly for Ramu, an enormous sea monster with antlers. I never found him. There are seventy-five levels of *Sea Hero Quest*. I was done by the tenth.

Released in 2016, the app was designed with a specific purpose in mind: to help dementia researchers study the earliest symptoms of Alzheimer's disease. In most cases, an initial Alzheimer's diagnosis still relies on a paper-and-pencil memory test. A doctor might ask a patient to remember a number sequence or draw an accurate image of a clock-face on paper. If she can't, this might be the first indication that the patient has dementia. Often, failure at this test marks the beginning of an unstoppable slide into confusion and memory loss—and eventually a complete loss of self. But memory loss isn't the first symptom. It comes later. In the beginning, there are other subtler processes taking place in the brain. Among the first neurons to die are cells that underpin the navigation system.

"One of the earliest things that goes wrong is that someone becomes disoriented in space," says Spiers. Suddenly, a familiar neighborhood becomes a bewildering maze. Researchers hope an app like *Sea Hero Quest* can provide a window into this stage of the disease. Perhaps the game could even provide an early detection system for dementia. For a researcher like Spiers, *Sea Hero Quest* serves a different purpose. He wants to know what the game can tell us about how healthy people navigate through space.

"We'd never really tested navigation on a mass scale at all," he says. "It's just too difficult." Within a month of its release, almost a million people had downloaded *Sea Hero Quest*. The moment they did, they became subjects in a study. "We've now got around four million people in the database to look at," says Spiers. "My interest here is: how do people vary all around the world and what patterns can we see in that?"

Somewhere on the outskirts of Cairo, a seventy-year-old man shields his cell phone from the harsh midday sun to steer the *Sea Hero* through a narrow inlet. In the northern world of rural Finland, near the Swedish border, a woman opens the app at night and plays when her children are asleep. An Australian teenager plays in the outback. In London, an office worker spends his lunch break cradling a sandwich and searching for Ramu, just like me. The amount of data available to Spiers is astonishing. Mountains of data. Players are located in all 195 countries in the world: in Canada, and Lebanon, and Malaysia, and Slovakia, and Switzerland, and everywhere else. They play in sprawling megacities and in remote unpopulated places, on the sides of mountains, in deserts, across the African continent, in Siberia, and on the thousands of tiny islands that make up Micronesia.

At the same time, Spiers and his coauthor Antoine Coutrot (a navigationally challenged 4) can crunch the data collected from millions of people at once. Published in 2018, the results are a revelation. One of his first findings, says Spiers, is a depressing one: it doesn't matter how well we navigate to begin with, our spatial skills are always worsening—and quickly. This surprised Spiers. He'd expected to see a sudden decline in old age.

"We just saw a continuous decline from the age of twenty," says Spiers. On a graph, the decline is a steep downward slope. It's Himalayan. If you're bad at navigating today, just wait until tomorrow. Comparing within a single age group, Spiers saw the spectrum of individual abilities that he'd expected, and that others have shown. But the shape of that graph is informative too, says Spiers. One might have expected spatial skills, when plotted on a curve, to be bell-shaped, or what's known as normally distributed, with most people clustered around an average performance, or the peak of the curve, and equal numbers of people in the tails of the curve who perform either very well, or very badly. Instead, the curve is lopsided. As Spiers puts it: "There's very few people who are really, really, really good, and there's a ton of people who are a little bit worse. But then the tail of the graph stretches out and out and out into all sorts of realms of badness."

There were more surprises. Looking broadly at the differences in spatial skills *between* countries, Spiers found that a player's country of origin determines how well she navigates. More specifically, the most important determinant of navigation abilities is the relative wealth of the country a player comes from—measured as per capita gross domestic product, or GDP. A player from Switzerland outperforms one from India; Singapore beats Colombia; Germany beats Iran, and Egypt, and the Philippines, and Serbia and Vietnam. And so on.

Navigationally, players from Scandinavian countries are rock stars. "Norway, Sweden, Finland, and Denmark and so on," says Spiers, "they all do very, very well and have a high GDP." What's more, he says, in Nordic countries, sports with a spatial component like orienteering* are popular, and the skills required to play them are taught in schools. The all-time medal table for the World Orienteering Championships reads as follows: Sweden (157), Norway (132), Switzerland (111), Finland (95). In fifth place is France, with a paltry 27 medals—almost all of which were won by one person, a wiry orienteering powerhouse called Thierry Gueorgiou who dominated the sport for the first decades of this century, winning everything.

When I asked Gueorgiou the same question I ask everyone, he, rather unsurprisingly, rated his spatial skills at 10 out of 10. Several countries performed slightly better at *Sea Hero Quest* than their GDP predicts—countries like the United States and Australia. In these countries, says Spiers, players are likelier to drive than use public transport, a fact that allows them the opportunity to develop their navigation abilities.

In the past several decades, a theory has emerged that men are better and more efficient navigators than women. Studies are published all the time that strengthen this narrative until it's become familiar to all of us—a kind of confirmation bias that props up the theory. We've all read carefully reported news stories with headlines like "Study: Men Are Better Navigators Than Women," and "Men Have a Better Sense of Direction Than Women, Study Says," and so on.

But it's not true, says Spiers. At least, it's not true in countries where men and women are treated equally. "In a place like Norway," says Spiers, "there's a negligible difference between men and women.

* Orienteering is a sport in which competitors use a map and compass to navigate from checkpoint to checkpoint across unfamiliar terrain at speed.

They just don't differ." That's because the gender gap between men and women is small in Norway, Sweden, and other countries where the sexes differ very little in their performance of navigation tasks. In countries with a larger gender gap, women experience lower levels of political participation and empowerment, economic opportunity, lower literacy rates and access to education. According to the World Bank's 2018 report "Women, Business and the Law," there are numerous countries with near-zero protection for women from violence, and few opportunities to build credit, or receive an education, or work outside the home. In almost forty countries globally, women cannot apply for a passport in the same way as men. Until as recently as June 2018, women in Saudi Arabia had been subject to a decades-old driving ban. How do you learn to navigate road systems if you're not even allowed to use them?

In these countries, the gap in navigation abilities between men and women is not negligible. It's a gulf. In Lebanon, men outperform women by a wide margin. There's no overlap. The same is true in Saudi Arabia, and Iran, and Jordan, and Egypt. In fact, it's true everywhere that women aren't free. It's striking. In other words, if you take away a woman's right to drive a car, prevent her from leaving her house, and carefully limit her activities when she does, you take away her ability to navigate too. Even so, factors like cultural differences, and GDP, and gender gaps can only explain so much, says Spiers. In any population—even a free and equal one—the spectrum in navigation abilities still exists. There will always be people who exist in that part of the graph that represents realms of badness.

"I think that part of the root of this is that people vary in how good their internal compass—an equivalent of a compass—and odometer and ability to memorize things is," says Spiers. "All these neural systems vary. People who are very good at judging spatial relationships in an environment had a very consistent sense of north in their brain—not the magnetic sense of north, but a sort of sense of the orientation of the environment. It's just cleaner—a clearer signal—in their brain than people who are bad at navigating. But we really don't know. It's a little bit of a mystery."

By themselves, place cells are inadequate. Even with their enigmatic and inscrutable firing patterns, and their fleeting glimpses into different possible futures, they're not enough to navigate by. When a discrete ensemble of place cells bursts into activity deep in my hippocampus, it encodes a specific location: the X that marks the spot. But there's a lot more to navigation than that. A place cell doesn't tell me what direction I'm moving in, where I'm planning to go, how far away my goal is, or whether I'm even traveling toward it. An ensemble of place cells firing in unison doesn't even tell me what direction I'm facing.

For that, my brain must rely on other neurons.

Chapter Four

The Perception of Doors

On January 15, 1984, James B. Ranck Jr. was lowering a recording electrode into a rat's brain in his lab at SUNY Downstate Medical Center in Brooklyn. It was a Sunday afternoon. Ranck was using the same techniques O'Keefe had been using in London more than a decade earlier. Back then, he'd even visited O'Keefe and Nadel in their cluttered lab that looked out over the traffic on Gower Street. Like them, he was recording the distinct firing patterns of single neurons to untangle the machinery of complex neural systems. He was trying to decode the brain, one neuron at a time.

"Jim Ranck decided to start looking at what happens to the signals once they leave the hippocampus," says Dartmouth University neuroscientist Jeffrey Taube (who rates himself a 9 out of 10). The hippocampus is divided into several distinct subregions, a little like the layers of a cinnamon roll. One of them, says Taube—known neuroanatomically as the CA1—is densely packed with the bodies of pyramidal cells, their long axons projecting outward to connect with neighboring parts of the brain. "The CA1 hippocampus is sort of the output area of the hippocampus." In other words, it's how the hippocampus communicates with other parts of the brain—and it has many connections. It projects first to a brain structure called the *subiculum*, which is Latin for "support."

It sits at the base of the hippocampus, like the ground floor of the structure that unfurls in an elegant S-shape above it. "Ranck's idea was: what happens downstream, in the subiculum?" says Taube.

It was a good idea. But Ranck missed his target.

In neuroanatomy textbooks, the brain is neatly separated into functional regions. Picture the diagram of a cow hanging on the wall of a butcher's shop, with the different cuts of meat clearly marked by dotted lines. Brisket. Flank. Sirloin. Anatomists have spent centuries diligently trying to determine the function and architecture of each different brain region—to find its borders, identifying where one region ends and another one begins.

In 1909, the German neurobiologist Korbinian Brodmann published a revolutionary work of brain mapping that is still used today. He systematically divided the cerebral cortex—the folded and ridged surface layer of the brain—into fifty-two distinct regions, based on their cellular architecture. His theory was that each of these areas (which became known as Brodmann areas) was distinct, with its own defined function. In some instances, he was later proven wrong. But, often, Brodmann was right. Even so, the process continues.

In November 2018, Greek-Australian neurocartographer George Paxinos announced the discovery of a previously unknown brain region. Located near the point where the spinal cord meets the brain, the region forms a distinct little island of cells in the middle of an information-carrying river known as the inferior cerebellar peduncle. Paxinos called the novel structure the *endorestiform nucleus*. Present in humans but absent both in our near relatives like rhesus monkeys as well as in more distantly related mammal species, its function is still unknown. But now it has a name. Discovering it, Paxinos said at the time, was like finding a new star.

On that Sunday afternoon in 1984, Ranck was aiming his recording electrode at a particular distant star: the subiculum. He hit its neighbor instead. Sitting just off target, the electrode eavesdropped on the neurons of the *postsubiculum*. Even though the two structures are just millimeters apart, they have distinct borders and demarcations, says

Taube. The distinctions between them are narrow but deep. Imagine boarding a flight to South Korea and landing unexpectedly in North Korea instead. Undeterred, Ranck listened in.

"That was when he discovered these cells that fired sort of like a compass," says Taube.

Years later, Ranck recalled the moment he found the first head-direction cell. Every time the rat pointed east, the cell fired. Just as O'Keefe had in London, Ranck was passing the voltage trace from the electrode through an amplifier and a speaker; when the cell fired, he heard a static-filled staccato *pop.* "That evening, I went to a party and floated on air," he wrote. "I told at least one friend at the party that I had recorded from a very exciting cell that afternoon. She did not have the slightest idea what I was talking about, but at least she was the first person I told about the finding."

He was floating on air. As long as the rat faced east: *pop pop pop pop.* When the rat turned to face another direction: *silence.* Ranck knew he'd discovered something remarkable. He brought magnets into the room to make sure the cells weren't somehow connected to Earth's magnetic field. It was a long shot. They weren't. "I put one rat on a twenty-foot cable and carried him around the room and into a new room, and the direction was maintained," he wrote. "The firing was correlated with absolute direction, independent of location."

It was startling to find that we all walk around with an internal compass—as unlikely as O'Keefe's and Nadel's place cell discovery years earlier. Later work has shown that, like rats, humans also have head-direction cells. If Paxinos finding the endorestiform nucleus was like discovering a new star, then Ranck's discovery of head-direction cells was like finding an Earth-like planet. It was like finding another Earth, teeming with cities and crisscrossed with highways.

•

Jeffrey Taube probably knows more about head-direction cells than anyone alive. He has spent more than thirty years trying to understand them—to untangle their function and meaning. In a photo of Ranck and

Taube taken in 1987, Ranck is a neuroscientist from central casting: he's slight, and nerdy, and he peers myopically through an oversized pair of black-framed glasses. His necktie, like a fat arrow pointing to the ground, is a '70s-era plaid specimen—an intermingling of browns, oranges, and creams. His breast pocket is lined with pens. By contrast, Taube is hip, in blue jeans and bearded, like an understudy for the Bee Gees, or a lesser known Bee Gee cousin. He cradles a rat in his left hand. In his right he grasps a long loop of cable, which, attached to an electrode on the rat's scalp, is recording the staccato firing patterns of the rat's head-direction cells.

At any single moment, most of a rat's head-direction cells are inactive. But a small population of them is firing as many as 200 times a second—and it's those cells that point the way, like the needle of a compass. Each head-direction cell has what Taube calls a preferred firing direction. These cells have the highest signal-to-noise ratio of almost all neurons in the brain. In other words, they're either on, signaling their preferred direction, or they're off, and silent. For instance, if I stand at the front door of my house and turn to face the street, I'm looking northward. The subpopulation of head-direction cells that is tuned to north begins firing—slowly at first as I turn toward north, but then faster and faster until they reach their maximal firing rate: I've hit north. If I keep turning, the subset of north-preferring head-direction cells gradually starts to quiet as another subset of head-direction cells with a different preferred direction begins to fire—as precisely as the needle of a compass.

"We call that a hill of activity," says Taube. If I turn on the spot, all of my head direction cells will be activated at one point or another during a single turn. They're scattered throughout a dozen or so interconnected brain regions: the postsubiculum, the anterior thalamus, the mammillary nuclei, in the retrosplenial cortex, and in parts of the brain stem—deep in the ancient innermost part of the brain that regulates basic survival functions like breathing and heart rate. Together, they form a brain-wide circuit. Even so, researchers struggle with some basic details, like how many head-direction cells there are in the brain. "It's impossible to know," says Taube, "because they're scattered throughout the whole brain. There's probably as many of them as place cells, if not more."

Switch off the lights, says Taube. Place the rat in total darkness. It doesn't matter. Its head-direction cells will continue to fire—for a while, at least. Even if some of the brain regions that contain head-direction cells are intentionally damaged with a lesion—like the hippocampal lesion that caused H.M.'s profound amnesia—the head-direction system never stops working, which is one of the benefits of a system located across regions.

•

The body is a compendium. It keeps a record of its own movements. And it's this internal record—known as *idiothetic information*—that provides the input for head-direction cells, along with spatial information gathered from visual landmarks. Idiothetic information comes from several sources, including *proprioception*, the awareness of the position of our body in space. For instance, even with our eyes closed, we know if we have our arms raised above our head. It also comes from something known as *efferent copy*: when the central nervous system sends a signal to the muscles, it keeps a copy of it so that it can refer later to the command—reference it afterward and make corrections to it if necessary. (Incidentally, efferent copy is the reason we cannot tickle ourselves.) It comes also via something called *optic flow*—a record of the visual flow of objects in the physical world as we move through it. Together, we integrate this information about our movements to help maintain the bearings of our neural compass. Precisely how we do it remains unknown.

In early studies, Taube placed rats in an environment with a visual landmark such as a white postcard, and recorded the preferred firing direction of a handful of head-direction cells. Then, he moved the landmark. The preferred firing direction of each head-direction cell moved by the same distance the landmark had moved. If the postcard was taken from the north-facing wall of the rat's environment and placed on the south-facing wall, a head-direction cell that previously had been tuned to north suddenly only fired when the rat pointed south: *pop pop pop pop*.

Some of the idiothetic information comes from our vestibular system—a complex of looping, fluid-filled canals nestled in the bone-locked labyrinth of the inner ear. They record movement too. The swirl of fluid in the tubes from head movements is detected by cells that bristle with microscopic hairs, thereby updating the brain on changes in head position. Little stone-like structures in the inner ear known as *otoliths* (from the Greek for "ear" and "stone") help to detect linear acceleration. All vertebrates, from humans to fish, rely on fluid-filled tubes and ear stones to regulate and monitor our movement through space. It's a strange and elegant system, until it stops working.

"Basically," says Taube, "if you disrupt the vestibular system, you disrupt the head-direction signal." A rat can no longer navigate as accurately if its vestibular system is damaged. And neither can a human.

•

Not long ago, I was lost in the woods in early autumn with a class of ten-year-olds on a school fieldtrip. Rain fell steadily between a thousand identical white pines, hissing between the branches. On the ground, pale ghostly mushrooms grew around the trunks of the trees, clustered like sea anemones at low tide. A thick, spongy carpet of leaf litter was slowly becoming soil. In a mossy clearing, kids were instructed to make temporary shelters from slippery, wet sticks. One of the boys wanted to go back to his cabin to get a rain jacket. I didn't know where we'd come from. He set off through the trees, making tight turns, first a left, over a little treeless rise, then a right turn and another left. It seemed completely random to me. The woods were crisscrossed with paths and non-paths. Without even thinking, the boy had constructed an accurate cognitive map like one of Tolman's rats and, with only identical-looking trees for landmarks, navigated in the rain through mossy undergrowth. I, on the other hand, had no inner map.

It's simultaneously funny but not funny: the clenched fist in the center of my mind, and the gargoyles of panic perched in the high branches of the trees, visible only to me. If there's one thing worse than being lost in the woods and crying softly onto your already-wet rain jacket, it's that a crowd of fifth graders might also be there to witness it. I tried to imagine my head-direction cells buzzing like little interconnected stars in my subiculum and elsewhere, firing as a single tightly organized neural unit across a dozen different brain structures. If place cells provide the melody—the specific individual notes and rests that, when combined, form a tune—my head-direction cells are the underscoring, providing an ever-present but barely perceptible harmony. But that counterpoint didn't seem to be present for me. Eventually, following a ten-year-old through the trees, I saw the cabin, wet-roofed in the rain. I knew where it was because I could see it.

But I didn't *feel* where it was like he did.

Years ago, I lived in a city in the American Midwest with a ring road around it. In total, the loop is eighty miles long, an endless curving four-lane highway, like a racetrack. A snake eating its own tail. One night, I needed to get on the ring road and drive north for a few miles and then exit—imagine driving on a clockface from three o'clock to two o'clock. Instead, I mistakenly began driving south, panicked, and drove the entire ring road, too disoriented and uncertain to turn the car around. I drove around the clockface. In the dark. It took hours. At different times in the journey, I drove in three different states.

My head-direction cells were never troubled.

•

"The people who don't have a good sense of direction usually rely much more on visual landmarks in order to navigate," says Taube. "People who have a good sense of direction, it's not like they can't use landmarks, but they have a better internal compass. If they close their eyes they can walk around better and keep track of where they're heading and where they are. My hunch is that they're utilizing their vestibular information better or in a different way than some other people."

Taube suspects these individual differences could account for the wide spectrum of navigation abilities found in any population. It explains why someone like Amanda Eller can lose herself on the trail and someone else might have a firmer sense of where they are in space.

"We talk about different intelligences," he says. "Someone could be a good artist, someone could be really good at music. I'm horrible at music. I can't carry a tune. I remember people, and what they look like, but I'm really poor at remembering their names. Obviously, there's something differently wired in peoples' brain that can then lead to different abilities. The sense of direction is just one of many different abilities we have."

New research suggests that head-direction cells might be more complicated than they seem, says Paul Dudchenko, a neurobiologist at the University of Stirling in Scotland. In fact, they don't even all do the same thing. "There's variability in head-direction cell properties in different brain areas, and even within a brain area," says Dudchenko, who gives himself a score of 6 and often gets lost playing *Call of Duty* with his son. "Sometimes the cells anticipate where the animal is going to be pointing its head, and sometimes they lag, and sometimes they anticipate by a slightly different amount."

Dudchenko thinks there might even be more than one kind of head-direction cell. In a 2017 study, University College London researcher Kate Jeffery (a 7 out of 10) discovered two distinct populations of head-direction cells in the retrosplenial cortex—a discrete region on the surface of the brain that seems to mediate between the brain areas involved in perceptual functions and navigation. "What she did is she simply had an animal walking from one room to another," says Dudchenko. The experiment wasn't that different from what Ranck did in 1984, picking a rat up and walking with it, tethered to a long cable, from room to room. "A traditional head-direction cell would just, whatever direction it's firing, keep firing that way as the animal moves into a new environment," says Dudchenko.

But not the cells Jeffery was recording in her lab at University College London. Once in a new environment, some of the cells maintained their preferred firing direction, just as expected. Others didn't. In the retrosplenial cortex, a particular subgroup of head-direction cells did something strange: they flipped. "As soon as the rat was placed in the other room, they switched direction," says Dudchenko. Tuned at first to one direction then—*flip*—they suddenly preferred another. "That should never, ever happen," he says. "In the traditional view, the head-direction cells would all maintain the same preferred direction as the animal moved from one room to the next."

Using electrodes, Jeffery began to look for the flexible direction-changing head-direction cells—she calls them bidirectional cells—in other brain areas. She only found them in the retrosplenial cortex. Jeffery thinks the flipping cells are solving what she calls a chicken-and-egg problem. Here's the chicken: among other inputs, head-direction cells rely heavily on visual landmarks to underpin our sense of direction. The egg: we need a functioning sense of direction in the first place to know whether a landmark is useful—to know if we should even bother including it in our internal maps at all. We don't want to build ourselves an error-laden map. In other words, in any new environment, even in a new room, most of a rat's head-direction cells stay focused on the global environment. They keep their bearing. North-preferring cells keep signaling north. But the bidirectional cells in the retrosplenial cortex focus on the new information. They reset.

Flip.

•

If the navigation system functions like an internal GPS, what does it need to be useful? First, it must include the precise locations of places or objects in space. This is the information encoded and cataloged by place cells in the hippocampus. Next, it needs a compass so the owner can know what direction she's facing, which is the work of the head-direction cells distributed in an interconnected network across the brain. But the brain's GPS still needs something else. A truly functional and informative map—a map that someone can navigate by and rely on—must include metric data too. It needs coordinates and information about distances. It needs a grid. In 2005, a Norwegian research team led by Edvard Moser (a 9) and May-Britt Moser (a 2 who became a 5)* announced that they had found the grid.

•

"In the mid-nineties," says Edvard Moser, "after twenty or twenty-five years of place cells, there was really still no consensus on how these cells came about." Nestled deep in the hippocampus, place cells are located far away from any sensory system—and that represented an unresolved puzzle to the Mosers. The hippocampus is as inaccessible as the ocean floor, so how does it understand what's going on at the ocean surface? "How can the hippocampus create a signal that is so precisely related to something in the outside world?" asked Moser.

In other words, place cells are well informed, but—as André Fenton had asked earlier—*how* do they become informed? Where do they get their information from? Neurons communicate with one another across a tiny gap that exists between them, known as the synapse. The Mosers—once married, now divorced, and still working closely together—decided instead to focus one synapse upstream of the place cells in the hippocampus, in a brain region known as the *entorhinal cortex*. Perhaps, they speculated, neurons there were feeding spatial information to the place cells in the hippocampus. Around 2002, at the Norwegian University of Science and Technology in Trondheim, the Mosers carefully aimed their recording electrodes at the little hills and valleys on the surface of the entorhinal cortex.

To find it, first turn a human brain over so that you're looking at the temporal lobes on its underside, says Menno Witter (6 out of 10), a Dutch researcher who collaborates often with the Mosers. On an upside-down brain, the temporal lobes form two curving elongated hillocks, one on each side of the brain, like parentheses. Witter, who I call around din-

* She says she began paying attention to landmarks.

nertime in Norway, says they look like two small sausages. Toward the center of the inverted brain, on its surface, sits the entorhinal cortex. It acts like a traffic hub, or a portal, says Witter.

"Essentially," he says, "it's the pivot that allows the hippocampus to communicate with the rest of the cortex." In either direction, both into and out of the hippocampus, information passes through the entorhinal cortex. If the hippocampus is a vast sprawling library, the entorhinal cortex is its front door. Smaller than a postage stamp, and slightly pear-shaped, for all of its spatial computing power, it's an incredibly diminutive structure—"maybe something like one-point-five centimeters square," says Witter. Even so, it's distinctive. "It's a very striking part of the brain that would be very easy to recognize if you know what you're looking for," he says. "It has little hills and valleys on the surface because it has clusters of cells that stand out like little bumps on the cortex."

Implanting recording electrodes in a rat's entorhinal cortex, the Mosers let animals move freely around an enclosed environment. That's when they found the grid cells: a population of neurons that fire in a highly organized grid-like pattern depending on the animal's location. The discovery—first published in 2005—earned the Mosers the Nobel Prize in Physiology or Medicine in 2014, sharing it with John O'Keefe for his work on place cells.

"These individual cells have a strictly hexagonal firing pattern," says Edvard Moser. "They fire in blobs that are patterned as equilateral triangles, or as regular hexagons, that repeat across the entire area where the animal is walking around."

Let's compare them with place cells: "There is one big difference between the grid cells in the entorhinal cortex and the place cells in the hippocampus," says Edvard. "In the hippocampus, the active place cells in any given environment are just a subset of all place cells. The rest are essentially silent. It's the combination of active cells that is quite unique for every possible environment, so that cells actually encode the environment and not only the location in the environment."

Not so for grid cells. They fire in multiple different locations—but their firing fields sit in a predictable geometric pattern. If two grid cells fire together in one environment, they fire together in all environments. Gradually, if an animal explores its environment, the entire population of grid cells is recruited, firing in an organized grid across the environment.

"What are they good for?" asks Edvard. "Well, they contain information about not only location but also distances and directions."

Like a precise coordinate system, the same geometric grid cell firing pattern exists whether I'm walking around my house, or through a busy grocery store, or navigating the pigeon-filled spaces of London's Trafalgar Square, or picking my way through dark and overgrown woodland at night. "If you provide inputs from grid cells to a computer," says Edvard, "the computer can immediately tell from the combination of inputs where the animal is but also what direction it's going in and how far it's going. This is information that doesn't really exist in place cells but is more strongly present in grid cells."

If place cells are discerning and specific—elegantly mapping the infinite with sparse coding—grid cells are no-nonsense generalists. They have one task, and they do it well. "Grid cells maintain essentially just one map that is used over and over again in every single environment, and that suggests that this entorhinal map actually doesn't care very much about the contents of the environment. It's more the structure, the metrics, the distance and directions within the environment. It's more a measurement system, in a way."

The location of grid cells, separate from place cells and one synapse upstream from them, makes structural sense to Moser. "Having one general measurement system and then letting this measurement system feed into the hippocampus and then be applied on every single map is probably quite functional," he says. "The hippocampus needs to distinguish places, so these hippocampal maps should be as different as possible in order to avoid mix-ups." Dividing it into two separate systems makes sense, provided the two systems connect to one another and work together. And they do, says Edvard Moser.

"It's a loop actually," he says. "It's not only grid cells, or entorhinal cells, feeding into place cells, but place cells feeding back to the grid cell system again. Since it's a loop, it probably means they reinforce each other all the time."

•

In a 2013 study, Columbia University researcher Joshua Jacobs (an 8) recorded grid cell activity in the brains of epileptic patients. He watched them firing in their distinctive geometric patterns as subjects navigated a virtual environment. "For a long time, people, including me, wanted to tell the simpler story that grid cells are the input to place cells—*the* input," says Jacobs. "And that's not right because place cells are active by themselves before grid cells even exist in young animals." Part of the problem in understanding grid cells, he says, is their location: they're difficult to reach.

"Measuring grid cells—individual cells—is hard because it's a deep brain structure," says Jacobs. "They're hard to record from, just for technical reasons. But an fMRI study a number of years ago, in 2010 by Neil Burgess, showed that you can measure grid cells by fMRI." By looking at them with an fMRI scanner, researchers are measuring grid cells at the voxel level—"which is probably half a million neurons," says Jacobs, "not one neuron at a time." Suddenly, the inaccessible has been made very accessible. Instead of listening to a lone grid cell at a time, it's possible to watch broader, more subtle changes taking place across a network of cells—fluctuating in real time. By monitoring many cells at once, neuroscientists can begin to understand the directional heading the grid cell network is representing—in other words, they can see the orientation of the grid laid over the environment by the brain.

"They had people imagine heading to different locations in an environment—close your eyes and imagine walking from A to B, A to C, A to E," says Jacobs, describing the 2010 study. "And, they found modulation just based on what direction you were *imagining* going in without even seeing exactly where you were on the screen or anything. It suggests that grid cells are really about more abstract representations than just low-level navigational processes alone."

•

Navigation is memory—impossible without the hippocampus, just like it was for H.M.—but it's much more than that. It's an act of perception too. Consider this: we detect only a small fraction, a sliver, of the physical world we inhabit, so our perception of it becomes very important. A dog has 300 million olfactory receptors compared to a human's six million, and essentially smells the world in rich three-dimensional detail. When my dog walks through the woods with his nose buried an inch deep in dead leaves, he's experiencing a version of the world I can never begin to know. He's walking through steep gradients of odor, bathing his olfactory membranes in a sea of molecules unknown and imperceptible to me.

A peregrine falcon can see light in the visible spectrum just as we do, but also in infrared. With its much larger lens, and a retina lined with densely packed cone cells, a falcon can also see an estimated eight times better than me, easily spotting prey from more than a mile away. It literally experiences its environment differently. For humans, navigation is largely a visual act. In any environment, our subconscious brain is furiously processing visual data—decoding space. We're hardwired to see landmarks. We see openings, pathways, routes, shortcuts, doors, and portals. Subconsciously, our brains register the subtle differences between a path and a blocked path. Memory might be the fuel for navigation, but visual information is the engine.

Russell Epstein studies the different brain regions involved in doing this work. "The very first experiment that I did, and this was twenty years ago, was to put people in an fMRI scanner and show them faces, objects, and also scenes—street scenes, pictures of rooms, landscapes," says Epstein, a cognitive neuroscientist at the University of Pennsylvania who rates his spatial skills at 7. "Much to my surprise at the time, we identified one region of the brain that responded very strongly and selectively to the scenes."

Epstein called it the *parahippocampal place area*, or PPA. If you show one of Epstein's subjects an image of a room, or a city street, or a landscape, the PPA goes into overdrive. It goes wild. It even processes the spatial details of a scene in people who have been blind since birth. Meanwhile, without a face to activate it, the nearby fusiform face area stays silent.

In a 2014 study, researchers implanted an electrode deep in the medial temporal lobe of an epileptic patient. With the electrode in place, they stimulated different brain regions directly with a small electrical current—regions like the PPA. Essentially, in the absence of any sensory information, the neurons were made to fire the way they do when someone sees a street, or the layout of a room, or a fast-food shop. And something incredible happened: the subject began to see places he wasn't in. He started to have spatial hallucinations.

RESEARCHER: Anything here? Do you feel anything, see anything?
SUBJECT: Yeah, I feel like I . . . *(looks perplexed, puts hand to forehead)* I feel like I saw, like, some other site, we were at the train station . . .
RESEARCHER: So it feels like you were in a subway?
SUBJECT: Yeah, outside the train station.
RESEARCHER: Let me know if you get any sensation like that again. Do you feel anything here? No?
SUBJECT: No, I . . . *(doesn't continue)*
RESEARCHER: Did you see the train station, or did you feel like you were in the train station?
SUBJECT: I saw it.

•

The PPA is not alone either, says Epstein. There are other brain regions that work to decode our surroundings and help us navigate. "It turns out that there are actually three areas of the brain involved: the parahippocampal place area, the retrosplenial complex, and the occipital place area," he says. Epstein thinks they work together as a functional interconnected network, with each of them playing a slightly different role in navigation.

"The parahippocampal place area and the occipital place area seem to be more involved in visual recognition of places and landmarks," he says, "and the retrosplenial region seems to be really crucial for using the visual information to figure out how you're oriented—sort of, how you're facing in the space."

•

Daniel D. Dilks, a researcher at Emory University in Atlanta, first described one of the three regions—the *occipital place area* (OPA)—in a 2013 study. It's located in a distant cortical region near the back of the brain, where visual information gets processed. For more than a decade, Dilks (who's a 4 or 5, and "not so good") has been studying the different brain regions involved in perception and spatial navigation—a domain he playfully calls Sceneworld.

"It's fine to know that you've got specialized bits of cortex," he says, "but the real interesting question is why? Why do we have multiple regions?"

By 2013, researchers had already described two components of Sceneworld: the parahippocampal place area, and the retrosplenial cortex. "I wanted to figure out the functions, the precise functions, of each of these regions," says Dilks. He did it with an fMRI scanner and an endless supply of willing human subjects. "When I started scanning people I noticed that there was another region that kept popping up across many of these people." It was near a brain area known as the *transverse occipital sulcus*—it's a deep furrow in the cortex that stretches from one hemisphere to the other near the back of the brain. It was activated by images of scenes. When he saw it lighting up again and again in one subject after another, Dilks wondered: could this be another scene-selective brain region?

To find out, he used a relatively new technique known as *transcranial magnetic stimulation*. Essentially, Dilks focuses pulses of magnetic energy directly at a particular brain region to disrupt its activity—like a temporary and reversible brain lesion. Before he can disrupt it, he must use an fMRI scanner to pinpoint the precise location of the OPA: since it's a functional region and not an anatomical region, he says, its location can differ slightly from one subject to another.

"You can't just for example look at the cortex and say, *That's going to be the parahippocampal place area because it's in and around the parahippocampal gyrus*," he says. "It may be, or it may not be." And you can't just aim a magnetic beam at the transverse occipital sulcus and assume it disrupts the OPA. Instead, he shows a subject in the scanner a series of different images—pictures of faces, scenes, and objects—mapping the region that sits near the transverse occipital sulcus and responds strongly to the images of places. Then he directs magnetic energy at that precise spot on the cortex. At the same time, Dilks asks subjects to make fine-grain distinctions between images of faces or houses.

"When I virtually lesioned that region in healthy individuals I could selectively impair their ability to recognize the places, but not the faces," he says. It was proof: another scene-selective brain region, along with the PPA and the retrosplenial cortex (RSC).

Through a series of ingenious experiments, Dilks has found a way to untangle the complex division of labor between the different brain structures in Sceneworld. "The common assumption was that all of these regions are involved in navigation," says Dilks, "but that had never been directly tested. It was just an assumption. I decided to ask, *Is that really true*?"

He began by placing subjects in an fMRI scanner and showing them images of different scenes. Then, he reversed the images: left became right, and right became left. First a beach, and then the mirror image of the beach.

"If all of these regions are involved in navigation—that's why we have them, because we have to get around our world—then they should all encode, for example, left and right information," says Dilks. "After all, when you're walking around the world you have to know if the table is to the left or right of you. You have to know if you're going to turn left or turn right."

Watching the neural activity expand and shrink on the cortical surface of the subjects' brains, Dilks saw differences begin to emerge. The OPA and the RSC both see the mirror images of a location as two different places. They detect the fundamental differences between a mirror image and its original. But the PPA—the parahippocampal place area—doesn't.

"Let's go back to our beach: the normal beach, if you will, and the mirror reverse," he says. "PPA saw it as the exact same place."

Dilks ran another fMRI experiment. "For you to get around your world not only do you need to know whether things are to the left or right of you," says Dilks, "but you have to know how far or near things are." This time, he showed subjects two images of an identical scene, but either close-up or from far away. Again, the PPA, which hadn't discriminated between an image and its mirror image, detected no differences between the two images either. "The PPA sees a picture of a house or a scene close to you and far from you as exactly the same place," says Dilks. As with mirror images, the OPA and RSC saw them as categorically different places.

A lot of work had already been done on understanding precisely how the brain recognizes objects. For instance, consider a cup. There are two totally separate systems in the brain for processing the image of the cup: one for recognizing the cup, and another separate system for reaching out to grab the cup. Recognition versus action. The information used by the two systems takes totally different paths through the brain, like trains traveling along two different tracks. Object recognition travels through the temporal lobe via what's known as the *ventral stream*, but the functional regions that govern action operate along a different route known as the *dorsal stream*, in the parietal lobe. Dilks wondered if there was a separation like this for the scene-processing parts of the brain too: "Perhaps, like objects, in Sceneworld we've got a recognition system and an action system." But, he says, "in Sceneworld, that's *recognizing* a scene and *navigating* a scene."

To see if this was the case, Dilks gave his subjects two different tasks while entombed in the scanner. "The person saw an image of a scene, and they either had to tell us what kind of place it was, or they had to imagine navigating through that place," says Dilks. These are two very different tasks. The image was identical. The only difference was what the subjects were expected to do with it. On the fMRI monitor, when subjects were asked to categorize the kind of place they were looking at, the PPA burst into life, unfurling like a colorful bloom on the surface of the brain. Meanwhile, near the back of the brain in the visual centers, the OPA was relatively quiet. When subjects were instead asked to imagine navigating the place in the photo, the OPA awakened.

"There are different kinds of navigation," says Dilks. It's not surprising that the OPA—located in the visual centers of the brain—assists with visually guided navigation.

•

Imagine, you enter a room, says Russell Epstein: immediately, the brain begins to process the spatial environment—to analyze and decode it. "There are many aspects of that scene that you might want to process," he says. "You might want to process the objects in the scene. You might want to process the overall shape of the environment. But an aspect of the spatial features of the scene that I think is really important is where you can go in the scene."

In other words, where are the doorways, the portals, the exits? Epstein calls them *affordances*. "If you're, as I am right now, in an interior room," he says, "you're probably very aware of where the door is: the place where you could egress the scene. Why is that different? How is it different from, say, a window, or a cabinet, or something like that?"

On a deep and subliminal level, the brain knows it is impossible to exit a room through a wardrobe, even though it looks similar to a door. This is also the work of the occipital place area, says Epstein. When we stand up to exit a room, the OPA is the reason we don't try to climb through a painting on the wall instead of through a door. It's why we don't try to exit a room via the refrigerator.

For a 2017 study, Epstein scanned the brains of subjects as they were shown fifty photographs of different indoor environments. During the scan, the subjects performed a task that was unrelated to finding the affordances in the room. For instance: *push a button if the scene is a bathroom*. As an array of different scenes—galleries, restaurants, kitchens, and bathrooms—flashed past, the subjects categorized them, and the fMRI scanner mapped the voxels of brain activity. Depending on the spatial structure of the scenes the subjects saw—and the location of the affordances in them—the activity in the OPA was different.

By the time he'd shown the subjects fifty different scenes, Epstein could begin working backward. From the activation patterns in the fMRI scans, he could reconstruct the spatial environment seen by a subject. By decoding the activity patterns in the OPA, he could predict certain spatial information about a room from a subject's brightly lit irregular voxels. Since subjects were asked to perform a non-navigational task—*push a button if the scene is a bathroom*—and not look for the affordances, Epstein says, the OPA must detect the affordances automatically. It breaks down a spatial environment subliminally—somewhere below thought.

•

The retrosplenial cortex has its own narrow role in spatial processing. "It's involved in what we call memory-guided or map-based navigation," says Dilks, "enabling you to represent where you are right now relative to where you live in a map way, or a memory-guided way."

Around 1900, Brodmann had noticed the RSC, calling it Brodmann areas 29 and 30. But for the next century, almost nothing was known about the structure—or what it did. It sits in prime spatial real estate. From there, it maintains connections with all the usual suspects—with the subiculum (head-direction cells), the hippocampus (place cells), the entorhinal cortex (grid cells), and the occipital cortex, where visual information is processed. In fMRI studies, the retrosplenial cortex responds to almost any navigation task you give it, particularly if the task involves the position of landmarks in an environment.

In a study carried out at University College London, Eleanor Maguire showed subjects lying in a scanner hundreds of photos of different objects—a bag of trash, a London bus, a lighthouse. She found that the retrosplenial cortex is more activated than other brain regions when subjects see an image of a permanent landmark. The retrosplenial cortex is awakened by an image of a building, but it ignores a bicycle. It is quiet when shown a bag of trash.

Next, Maguire compared good and poor navigators. In subjects who had given themselves a high score on the Santa Barbara Sense of Direction Scale, the RSC is even more activated by permanent landmarks. It seems that the retrosplenial cortex is involved in mapping them. Next, Maguire asked the subjects to rate how often they expected the objects in the images to move in everyday life. *Give them a number from 1 to 5*, she asked the subjects. A rating of 1 meant an object moves very often (think London bus, or bicycle); a 5 was given for objects that never move (think lighthouse, or mailbox). Good and poor navigators didn't disagree when scoring the impermanent and moveable objects. Everyone agreed that a London bus moves very often, and so does a bicycle. But poor navigators failed to reach a consensus when they rated the permanent objects—the 5s. If you show poor navigators an image of a fire hydrant, or a windmill, or a wooden cabin, they will tell you it moves.

Their retrosplenial cortex has failed them.

The poor navigators in Maguire's study aren't unique. One man, whom scientists described in a case study, could pinpoint the exact moment his retrosplenial cortex stopped working: it was the evening of December 11, 2000. The streets were dark. The man, a fifty-five-year-old taxicab driver, was driving home after a long shift navigating the crowded streets of Kyoto in Japan, when, suddenly, he forgot how to get home. Hands on the steering wheel, he peered up at the city blocks as they slid past the cab, scrutinizing the little checkerboard squares of lit and unlit office windows in the darkness. He realized he could recognize the buildings. He understood where he was. But that information was useless. He couldn't convert that spatial knowledge into the information that would get him home. He had no sense of direction.

It was later determined that the man had suffered a brain bleed on his retrosplenial cortex. There, on the grooved surface of the brain, a vessel had ruptured its thin walls, like a river bursting its banks. He felt no pain and experienced no other symptoms. In the end, he called his wife, and she guided him home, street by street. His egocentric spatial representation—the ability to determine where he was from his own position in space—had been damaged. But his *allocentric* sense of position—that is, his map-based position based on features in the environment—had been disrupted too. He could point correctly to cities on a detailed map of Japan, and find Australia on a world map, but he couldn't draw an accurate floor plan of his own house.

There are other examples in the scientific literature, most of them involving damage to the right retrosplenial region. I have begun to collect them, just as I do the news stories of lost people. In another case, one man showed a superlative degree of lostness, also after suffering an infarction in the right retrosplenial cortex: he "suddenly became unable to return to his room from a toilet that was about 20m away." A single burst blood vessel, and his bedroom was completely unreachable.

This is what happens when a brain region packed with head-direction cells and other spatial neurons, and closely connected with the hippocampus, suddenly stops working.

•

"The kind of work that I've been doing at least so far has been mostly fMRI studies in the normal population," says Russell Epstein. For decades, he's been trying to understand how the normal human brain functions. But not anymore. Epstein says he wants to understand what makes some people struggle more than others during navigation tasks. One way to understand that is by looking instead at brains that aren't working—at atypical brains.

In many cases, an injury to a functional brain region, from a stroke or a traumatic brain injury, can cause what's known as an *agnosia*—from the Ancient Greek for "to not know." There are all kinds of agnosias. Each one represents a strange misfiring or malfunction of the brain that, depending on the damaged structure, causes a very specific kind of not-knowing. Importantly, an agnosia can provide a window into the function of different brain regions. Taken as a group of disorders, agnosias are an important reminder of just how many different kinds of information we subconsciously sift through to make sense of the world.

For instance, someone with phonagnosia is unable to recognize familiar voices, even though she understands what is being said. Finger agnosia prevents sufferers from distinguishing one finger from another: their map of *themselves* is disrupted. Other people suffer from category-specific agnosias. In other words, they are unable to recognize certain types of objects only. For example, a patient might be unable to tell the difference between living and nonliving objects. In a 2002 case study, a subject known as DW was unable to identify vegetables—only vegetables. People with time agnosia are unable to determine the duration and sequence of events—their understanding of time has unraveled. Damage to the fusiform face area, the narrow band of the cortex involved in facial recognition, causes prosopagnosia, or face blindness. Sufferers are unable to recognize faces—even their own children. In extreme cases, prosopagnosics are strangers even to themselves. When they look into a mirror, a surprised but unknown face looks back at them.

Agnosias have been incredibly important to understanding and confirming the location of different brain regions involved in navigation—like the retrosplenial cortex. If the cab driver in Kyoto had suffered an injury to the parahippocampal place area, says Epstein, he would no longer have recognized the buildings and city streets as he drove past them in the darkness.

"This would be someone who can recognize faces, and maybe objects—particularly small objects. But not buildings," he says. In one case study, a seventy-one-year-old Canadian woman suffered damage both to the fusiform face area and the parahippocampal place area. Suddenly, she began to suffer what are known as *delusional misidentification syndromes*, believing her husband was her dead sister and her house was a perfect rented replica of her real house. Sometimes, she would pack her bags so that she could return to her real home and leave the strange replica house behind.

•

Other sensory systems are also involved, and in ways we still don't fully understand. In 2018, scientists at McGill University in Montreal tested the hypothesis that olfaction—our sense of smell—evolved partly to assist with navigation. Evolutionary biologists have reasoned that, since most animals use chemical cues to find food or avoid predation, perhaps the two processes of olfaction and navigation evolved together, informing and reinforcing one another. First, psychologist Véronique Bohbot tested the ability of a group of volunteers to smell forty different odors. One after another, subjects raised felt-tip pens impregnated with odors to their noses and inhaled deeply: ginger, cigarettes, grass, lime, garlic. Then, she placed them in a virtual environment. Bohbot found something remarkable: the individuals who performed best at the smell test were also far better at navigating through a virtual town. What's more, the link between olfaction and navigation was strongest in subjects whose left medial orbitofrontal cortex was thicker, and whose right hippocampus was larger in volume—two of the brain regions involved both in spatial memory and olfaction.

Most likely, there are lots of other specialized neurons and brain regions we haven't even discovered yet that are also involved in spatial tasks. They're probably hidden away in deep brain structures, performing implausibly specific jobs. In the past decade or so, researchers have described some of them. "We have a cell that codes for directionality," says entorhinal cortex expert Menno Witter. "There are cells that code for speed. There are cells that code for borders—for boundaries in the environment." In July 2019, Jeffrey Taube published a study of the postrhinal cortex, a brain region close to the PPA. From the specific firing patterns of the cells there, Taube suspects the region might be involved in combining egocentric and allocentric spatial information.

That's not a trivial task: "When spatial information enters your brain, it has to enter it via an egocentric system—meaning you see things in the real world based on how you're looking at it," says Taube. But then the brain has to repackage that egocentric information. It must transform it into an allocentric reference frame—one based instead on the location of features in the world with respect to each other and not based on your place in it.

For instance, from the top of the London Eye—the enormous observation wheel that slowly turns on the South Bank of the River Thames—I can look southwest across a serpentine bend in the Thames and pick out the Palace of Westminster and the clockface of Big Ben. Farther west, beyond the green wedge of St. James Park, is Buckingham Palace, solid and biscuit-colored on its apron of concrete. At almost five hundred feet in the air inside the silver egg-shaped passenger capsule, I can see their locations. But that doesn't really help me later when I attempt to navigate the mile-long walk from Big Ben on Parliament Street, west along Birdcage Walk to Buckingham Palace. For that, I need to convert the egocentric information to an allocentric reference frame and use it to construct a map: a complex event that happens somewhere in the brain. "That's sort of like the sixty-four-million-dollar question: where does it happen?" says Taube. He thinks it might take place in the postrhinal cortex. Like so many other components of the spatial processing system, it was found by accident—by a student who was trying to listen in on cells in the entorhinal cortex and began recording from a different brain region instead.

"In order to stick electrodes in the entorhinal cortex," says Taube, "the electrodes first have to pass through the postrhinal cortex." On the way, with the electrode lodged in the postrhinal cortex, the student paused for a few moments, turning on the apparatus to record cellular activity there. "It was a little bit serendipitous," says Taube. If it hadn't happened—that little moment of serendipity—who knows when someone might have discovered the postrhinal cortex does this too?

Like most of the work that involves eavesdropping on single cells, Taube's student was working with rats. In humans, the postrhinal cortex is known as—drumroll please—the parahippocampal cortex, a region that neuroscientists already know is important for navigation. A question remains: exactly how does the parahippocampal cortex contribute to our ability to navigate? University of Freiburg researcher Lukas Kunz thinks he might know the answer. All the other spatial cell types—O'Keefe's place cells, Ranck's head-direction cells, the Mosers' grid cells—encode space in an allocentric reference frame. External space. As a cognitive map. But somewhere in the brain, there must be another kind of cell that tells us about the egocentric (or self-centered) relationship between us and the external world—not the map itself, but where we *are* on the map. When we navigate or remember previous experiences, we do so from a first-person, egocentric perspective.

Kunz was recording single cells in the brains of epilepsy patients when he found neurons that perfectly fit that activity profile. He's tentatively calling them anchor cells. But he's still cautious about what they're showing, and the findings need to be replicated by other researchers. Even so, the cells seem to be providing that egocentric information. Imagine that you're walking home, says Kunz: specific anchor cells become activated when your home—the *anchor* in this scenario—is ahead of you and several hundred meters away. Other anchor cells activate when your home is to the left of you and only a few meters away. Their firing patterns are determined by your position relative to the anchor. Instead of place cells, which endlessly signal world-based location data—*this is your front door; this is Times Square; this is Stonehenge*—Kunz thinks ensembles of anchor cells instead tell you where you are in relation to those locations. It's the sixty-four-million-dollar question. Kunz might have answered it.

Every day—in fact, hundreds of times a day—different brain regions dance a light-footed waltz with one another, seamlessly passing spatial and sensory information back and forth, interpreting and refining it as it moves across the brain. The waltz often becomes most obvious when it suddenly begins to break down. A dancing partner is absent. Another forgets his steps, or treads on his partner's foot, or leads when she should follow.

Think of the endless ways we use these different neural systems:

when we navigate the confusion of a busy airport; or when we walk from our favorite chair to the bathroom; or when we take an impromptu detour to bypass a congested road. They're constantly waltzing, producing a spatial representation both of the world and *our place in it*. What's more, they've been doing it for a very long time.

Chapter Five

The Obligate Symbolists

At five-foot-one, the man is smaller than me, barrel-chested and sturdy. His shoulder-length hair is dark and tangled like a mane, and so is his untrimmed beard. With his chin tucked into his chest, he looks at me quizzically from beneath his brow, a protruding ridgeline that casts a shadow over his eyes. His nose is wide—researchers say its shape helps to humidify and warm the dry icy air of his northern environment.

He's tattooed: two slate-grey chevrons begin at his shoulders, slanting downward, almost-blue beneath the thatch on his chest, to terminate just above his sternum. Mostly, the man looks a lot like me—a bit wild, perhaps. But he's not me. Technically, he's not even a man.

He's a Neanderthal: *Homo neanderthalensis* to my *Homo sapiens*. He stands in a glass display box in the Human Evolution gallery at the Natural History Museum in London beneath a square of bright white light. It's as if he's fallen through a portal, tumbling for 40,000 years into a darkened room in South Kensington filled with running schoolchildren, their backpacks bouncing on their shoulders. He's based on Spy II, an almost complete Neanderthal skeleton discovered in a sprawling multichambered cave complex in central Belgium in 1886. It's worth mentioning that I had to make several attempts to find the Neanderthal in his box, bathed in his square of unnatural light. The exhibition halls were darkly lit and confusing. Several times I had walked into little alcoves and dead ends filled with rows of skulls mounted on little pedestals. More than once, my search ended at the very non-Paleolithic gift shop, as if I had been funneled there.

But I eventually found him.

He's the creation of Dutch reconstructive artists Alfons and Adrie Kennis, who spent hours painstakingly converting his skeletal remains into a life-size model of a male Neanderthal in his twenties. He was put here in a box so that modern English schoolchildren could point at his nakedness and laugh. But when he was alive, how did Spy II see and understand the world? Did he see it the way I see it? How did he navigate it?

•

Forty thousand years ago, a battle was raging across what is now mostly Europe. It was a battle for existence. For hundreds of thousands of years, Neanderthals had owned the wild places, dominating the temperate woodlands of Portugal and Spain and, to the north, the tight mountain passes through the Swiss Alps. Their range stretched in a band from Wales and England in the west, eastward across central Europe and the Balkans, all the way to the Altai mountains in Siberia. But something new was arriving, something very dangerous. Advancing rapidly toward them, surging northward from Africa in a cloud of dust, the new threat was us: modern humans. Within a few thousand years of our arrival, Neanderthals were gone forever. Erased. The speed of it—its suddenness—was staggering. For hundreds of thousands of years, Neanderthals had been part of the landscape.

"All we know is that *Homo sapiens* showed up and Neanderthals disappeared," says American Museum of Natural History curator Ian Tattersall (a 6 or 7). "They'd been through very hard times before. They'd been through a lot of climatic and environmental vicissitudes, and the only thing that was truly new in their experience was *Homo sapiens*."

We ended up being too much. We're implicated in their disappearance, but it's not clear how. The evidence is circumstantial. Now only the bones remain—skeletons like Spy II in Belgium, and others unearthed at Neanderthal sites as far-flung as France, Turkey, Israel, and Uzbekistan.*

•

If we rewind 500,000 years to the Middle Pleistocene—reversing every evolutionary adaptation, one by one—we finally return to the moment when Neanderthals and modern humans diverged from *Homo heidelbergensis*, their common ancestor. *Homo heidelbergensis* was intrepid. At some point, a few left Africa. They dispersed, radiating outward to Europe and beyond. Their bones have been found at fossil sites as far afield as Germany, Spain, Greece, and France.

Once in new environments, they evolved until eventually they became different enough to be their own distinct species: the Neanderthals. Other members of *Homo heidelbergensis* stayed in Africa. Around 300,000 years ago, anthropologists believe they evolved into modern humans. That's when we began. Eventually, some of us started leaving Africa too. Around 55,000 years ago, we surged northward from the fertile floodplains of the Okavango Basin in present-day northern Botswana, to colonize every part of the planet.†

* Today, anthropologists think Neanderthals were more like us than unlike us. They probably possessed language but perhaps it was not as sophisticated as ours. Their brains were very similar to ours. The reasons for their disappearance are complex and probably also involved factors like group sizes and population dynamics, competition with humans for resources, and the effects of rapid changes in climate.

† This is the likeliest sequence of events, but it's still only an informed guess. The fossil record is incomplete; in places, the gaps in it are hundreds of thousands of years long. Anthropologists debate the timeline and the confusing taxonomy of a shifting array of early hominin species: *Homo floresiensis*, discovered in 2004 in Indonesia and alive 18,000 years ago, probably a descendant of *Homo erectus*; *Australopithecus sediba*, found in 2010, a 2,000,000-year-old fossil similar to its contemporary, *Australopithecus africanus*; the Denisovans, another extinct human species, known from just a few teeth and bone fragments found in 2010 in Denisova Cave, in south-central Siberia.

It's also worth mentioning that if early modern humans had spatial abilities as poor as mine, we might never have left Africa. How would we? There's evidence that modern humans had left Africa before—but faltered. In other words, earlier attempts to leave Africa failed. In 2018, scientists announced the discovery of a human fossil that dates back to around 180,000 years ago: eight intact teeth in a curved fragment of jawbone excavated from the floor of a collapsed cave in Israel. In July 2019, scientists announced the discovery of an even older human fossil—this time in Greece. Piecing together a jigsaw of bone fragments, the researchers rebuilt the skull of Apidima 1, a human skull. It's around 210,000 years old.

From the moment humans came into existence, we were mobile navigators. So, what set us apart?

•

By the time we collided with Neanderthals, we were categorically different. The world hadn't seen anything like us before. We had invented gods and rituals. Abstractions. Superstitions. Ghosts. Everything had meaning beyond its immediate meaning. In other words, we had begun to process the world symbolically. Modern humans, Tattersall says, were "drenched in symbolism." The evidence of it is everywhere in the archeological record—in cave art, body decorations, musical instruments, ritual burial, and sculpture. When I corresponded with University of Liverpool anthropologist Laura Buck, she described us as *obligate symbolists*. In other words, we can't *not* indulge in symbolic thought. It defines us.

For decades, researchers thought the earliest musical instrument was a Neanderthal artifact—a 43,000-year-old flute made from a cave bear femur, discovered at a Neanderthal site in northwestern Slovenia in 1995. It probably isn't. Now most anthropologists accept that the object was likely made by hyenas chewing inadvertently on the bone, and that the holes are ancient toothmarks. The object now thought to be the oldest known instrument is a long and slender bone punctuated by five perfectly round finger holes. The holes show the careful and deliberate signs of tools. Made from the bone of a griffon vulture, the flute was discovered in 2008 in a hillside cave in southwestern Germany, a site littered with the leftover debris of life—human life. Using a carbon dating method, scientists date it to around 43,000 years ago. The discovery suggests that modern humans brought music with them as they entered modern-day Germany via the Danube Corridor.

This difference between us and Neanderthals is important: our symbolism is uniquely human.

"I'm not seeing the Neanderthals as being routinely symbolic," says Tattersall. "I think Neanderthals were extremely complex—don't get me wrong. I have plenty of admiration for them. They were resourceful. They were flexible. They were clever. But I think what they show is that you can be clever without being clever in exactly the same way that we are."

Humans tend to think that the only way to be clever is in our own stereotypically human way, says Tattersall. By that reductive thinking, our closest relatives were just slightly less efficient versions of us.

"I think that's the wrong perspective," he says. "Neanderthals were behaving in their own way, and they had their own form of relationship to information and to the environment. We're very wrong to try to interpret them in our own terms."*

•

Neanderthals moved within a limited territory and maintained small, manageable groups. But humans maintained sprawling complex relationships with more people, across larger territories. And that's impor-

* Anthropologists debate whether Neanderthals processed the world symbolically. Even if they did, the evidence for it is nothing like the record showing human symbolic thought. For instance, Neanderthals might have buried their dead like we did, but maybe they did it simply because dead bodies smell and not for superstitious reasons. Neanderthals might have adorned their bodies like we did, but maybe they were just emulating humans, a new arrival to their territory. Neanderthals possibly painted cave walls, but the instances are rare, and archeologists suspect some of the most artful aspects of that work might have been added later by humans. It's possible that Neanderthals were in the process of developing symbolic thought when they disappeared.

tant, says Ariane Burke (8 out of 10), an anthropologist at the University of Montreal in Canada.

In fact, Burke authored a paper in 2012 that suggests it made all the difference. "Spatial cognition is tied to things like managing a social network that's spatially extensive and that's dynamic," says Burke. The far-flung social connections maintained by humans required additional computing power, she says. These connections were another abstraction; they required more symbolic processing. "You have to try to remember *where* people are located and *when* they might be found in certain places," says Burke, "and then you have to figure out how to maintain contact with them, either physically or through proxies."

Brains are plastic. They're altered by what we ask them to do. Since spatial cognition is a use-it-or-lose-it kind of activity, when humans adopted these abstractions, they also began to develop more advanced spatial cognition. This changed their brains, widening the cognitive gap between humans and Neanderthals.

Burke doesn't want to oversell these ideas—Neanderthal and modern human brains are incredibly similar, she says. Even so, the concept that the brain is changed by the activities we ask it to perform should sound familiar. Think again of bus drivers and cabbies in central London. After all, their brains are even more similar to one another than those of Neanderthals and humans, but University College London researcher Eleanor Maguire has seen enormous differences in the activity levels of certain brain regions. A taxi driver can hold the complex geography of the entire city of London in his head—and he has a measurably larger hippocampus to show for it. Driving the same loop over and over leaves a London bus driver with a smaller hippocampus. In her paper, Burke writes: "In other words, the plasticity of the human brain allows significant differences in spatial abilities (and neuroanatomical changes) to develop when a subject is provided with training opportunities."

Flexible and highly mobile—the obligate symbolists—we modern humans moved around a lot, expanding our ranges, maintaining complex sprawling abstracted social networks. We could hold the complex geography of an entire landscape in our heads.

Modern humans can study to become London cab drivers.

•

Anthropologists like Burke face what seems to be an insurmountable problem: 40,000-year-old brains don't stick around. Since brains are made of soft tissue, they don't fossilize like bones do. Encased in the skull, they just disappear before they can be fossilized. When they're gone, the neuroanatomical secrets they once contained are gone too. Modern humans and Neanderthals were categorically different from one another, and those differences would have been reflected subtly—and perhaps less subtly—in the specific ways their brains functioned. But one species survived, and the other went extinct. Thousands of years later, we have no way to understand the differences between a human and Neanderthal brain.

But Ralph Holloway does. In his lab at Colombia University, Holloway (a 7) is cradling what looks like a misshapen bowling ball in his hands. He's a paleoneurologist. Behind him there's a maelstrom of partly scientific clutter: crammed bookshelves and leaning towers of binders; rolls of duct tape; two fully articulated replicas of the human skeleton; an asthma inhaler, energy bars, a tape measure. In an unvisited corner of the room, the leaves of a potted plant are wilting and turning brown. A long worktable is littered with skulls, and half-skulls, and fragments of skulls, the pieces like cracked and broken pottery.

Holloway turns the ball over, inspecting its hollows and curves and convolutions. It's blue—the unalloyed blue of a cloudless September afternoon. A Greek sea. On its underside is a thick stem, as if it were a flower and this was its stalk. Looking more closely at the blue ball, its surface is marked by the riverine meanderings of cranial nerves and blood vessels. It's bursting with information—data. The object Holloway is holding is an *endocast*, a three-dimensional cast of the interior of a skull.

Under certain conditions, endocasts form naturally. Forty thousand years ago, when a Neanderthal like Spy II died, his body was covered with a layer of sediment, which slowly filled his cranial cavity, and then lithified into solid rock. In other words, it became a brain-shaped boulder—a brain-shaped boulder in the rubble of a hillside dump near Johannesburg in 1966, or a brain-shaped boulder that rolled to the surface of a travertine quarry in Slovakia in 1926 like a misshapen cannonball. Surrounded by the fossilized bones of forest elephants, it's an estimated 105,000 years old.

Discovered in August 2008 on a construction site in Heslington, Yorkshire, the oldest existing brain is around 2,600 years old. The color of an over-baked cookie, the brain is furrowed with intact cortical folds and ridges. It's an anomaly—preserved by the oxygen-poor conditions of the Heslington mud. In January 2020, scientists announced the discovery of another ancient brain. It looks like the sort of object you might see resting on a stand in an upscale boutique in Paris: an irregular jagged piece of black glass with serrated edges. Scientists say it's a vitrified human brain. It was removed from the cranial cavity of a body found at Herculaneum, near Pompeii—a body instantly reduced to ashes during the cataclysmic eruption of Mount Vesuvius in 79 CE.

Neither the Heslington brain nor the glass brain from Pompeii is very informative—at least, not like Holloway's endocasts.

Holloway makes them himself, carefully injecting molten latex rubber into the cranial cavity of a skull, like pouring Jell-O into a Jell-O mold and waiting for it to set. The stalk on its underside is the brain stem as it tapers into the spinal cord at the base of the brain and departs the skull. Pulsating and throbbing in life, and overlaid with a web of blood vessels, a brain leaves an impression on the interior walls of the skull. An endocast is its negative. In some cases, an endocast includes the contours of specific ridges on the surface of the brain. It's not the same thing as a brain—not even close. But it's a deeply informative artifact, and can be studied much like the characteristics of an object can be studied from the shadow it casts.

"I have three or four hundred of them," says Holloway. "About two hundred of them, at least, are from the apes: gorilla, bonobo, chimpanzee, orangutan, and quite a number of the Hylobatids—the gibbons. Those were all made from museum specimens, and they're in latex rubber, or some other kind of thing." Since an endocast reflects all the unseen subtleties of the inside of the skull, Holloway can use them to understand the differences in cranial volume and the relative sizes of specific brain regions in different species.

On another table, there is an orderly row of human skulls. In front of each skull, upside down like a turtle turned on its shell, is an endocast. The casts are aquamarine, and turquoise, and lilac, and sage green. "I did about eighty modern human skulls," says Holloway. He stood, legs slightly bent, swirling the skull like a punchbowl to coat its inner surfaces with molten latex rubber. Once dry, Holloway collapses the rubber, then retrieves it, a millimeter at a time, from the foramen magnum, the hole at the base of the skull, like a magician pulling a handkerchief from a closed fist.

Then there are the older skulls. "The fossil endocasts probably number around 100 to 200," says Holloway. Some of them are small and irregular, and were made using a few fragments of incomplete fossil, but others are complete—as marbled with cranial nerves and blood vessels as an endocast from a modern human skull. Holloway has been making endocasts since the 1960s, working with some of the most important human fossils ever found. He made endocasts alongside Richard Leaky in Nairobi. In Frankfurt, in the middle of a superlatively hot summer, he sat at his workbench in his underwear, injecting latex into fossil skulls excavated from the Solo River in Java.

Today, paleoneurologists rely mostly on virtual endocasts—perfect high-resolution three-dimensional images of the interior structure of a cranium, obtained with a CT scanner. Even so, preferring a physical object over an image of an object, Holloway will 3D-print a virtual endocast so that he can hold it in his hands. He can feel its contours and subtleties with his fingertips. If making endocasts is an art form—and it is—then Ralph Holloway, who is eighty-four years old and now uses a walking stick to move around his lab, is its Picasso.

What do his endocasts tell us?

First, our brains are smaller today than they were in the Paleolithic era. For tens of thousands of years, they've steadily been shrinking.

"We have smaller brains than Neanderthals did," says Ian Tattersall, "but our ancestors—our *Homo sapiens* ancestors who were contemporary with the Neanderthals—had brains as big as Neanderthals. Sometime since the late Ice Age our brains have in fact declined in average size." Most likely, this is because we process information differently from Neanderthals and early *Homo sapiens*. "Our way of doing it today is more energetically frugal and thereby advantageous," says Tattersall. "We don't need the amount of brain that a Neanderthal or an early *Homo sapiens* had."

But Holloway's endocasts tell us much more than that. Over millions of years, they show a pattern in the changes that took place in hominin brains: first, there's a reorganization of brain structures, followed by a sudden increase in brain size. Reorganize . . . expand . . . repeat. Each cycle takes thousands of years.

When we compare human and Neanderthal endocasts, other important differences become obvious too. Placed side by side, an endocast of a modern human skull looks very different from a Neanderthal endocast. A human braincase is large and rounded, and a Neanderthal's is flatter and elongated. For a 2018 study, Simon Neubauer (6 or 7), a researcher at the Max Planck Institute for Evolutionary Anthropology in Leipzig, wanted to know precisely when the human brain gained its characteristic shape. He began by making virtual endocasts of fossilized *Homo sapiens* skulls that represent different geologic time points. Some of them are among the oldest known human fossils, dating back to around 300,000 years ago. They are examples of the earliest fossil remains that anthropologists still consider anatomically modern humans. Others are more recent: a 120,000-year-old, mostly intact skull from Tanzania, discovered by Mary Leakey in 1976; two 14,000-year-old skulls from Bonn-Oberkassel in Germany found in 1914, buried with the bones of a domesticated dog; and a skull unearthed by a laborer in 1909 in Combe-Capelle in southwestern France—an estimated 8,000 years old. Neubauer compared them with endocasts from Neanderthals like Spy II, and *Homo erectus* specimens from Kenya and Tanzania, along with a few complete *Homo heidelbergensis* craniums, and the skulls of almost a hundred present-day humans. He found that the change from an elongated braincase to a more rounded modern shape was gradual, and that he could see it happening across time. For instance, an endocast of Qafzeh 6, a human skull found in Israel in 1933, is an estimated 115,000 years old. It's clearly human, but not yet fully modern.

Between 100,000 and 35,000 years ago, there's a gap in the fossil record, says Neubauer. And that's important. What happened during the gap? At the start of it, humans didn't yet have their large, rounded, modern brain shape. By the end of it, they did. It was sudden. At some point between 100,000 and 35,000 years ago—within that frustrating fossil gap—we gained our modern brain shape and simultaneously came roaring out of Africa, dispersing across the world to occupy every corner of the planet.

Perhaps that's a coincidence. But maybe it's not.

•

Over time, as the human brain reorganized and grew, some brain structures changed more than others. This is how we became modern humans: the symbolists. In developing a larger, rounder shape, the human brain suddenly expanded in one region, located slightly toward the back of the brain and known as the *parietal lobe*—the parietal cortex makes up the surface of the parietal lobe. Originally, one of the main functions of the parietal lobe was to process sensory information about our body and its position in space. In the clinic, neurologists can assess parietal lobe function by asking a patient to make a ring shape with the thumb and index finger of one hand—the okay sign—and interlock it with the same shape made by the other hand. Someone with a parietal lobe injury wouldn't be able to do it.

But the parietal lobe is involved in a lot of other seemingly unrelated cognitive functions too, like understanding numbers, time, processing feelings of empathy and forgiveness, and even happiness.

At some point, though, says University of Colorado paleopsychologist Fred Coolidge (8 out of 10), an *exaptation* occurred. An exaptation is a physical trait or feature that takes on a function that differs from its original function. For example, feathers first evolved in non-avian dinosaur species as a means of keeping warm. But later, feathers exapted and became used for flight instead.

In the same way, says Coolidge, when the parietal lobe expanded in size, it enhanced our visual and spatial skills. But it exapted too, allowing complex functions like numerosity, and sophisticated human emotions like empathy—and navigation.

As we navigate, combining *egocentric*—self-centered—spatial information with *allocentric*—world-centered—information, the parietal lobe becomes very busy. Since this brain region is involved in tracking the location of the body and its parts, it makes a sense that it would do some of this work. More specifically, neurobiology studies of modern brains have shown that the parietal cortex encodes egocentric space. When you're dreaming, or reexperiencing a spatial memory, or imagining the layout of a familiar room, the neurons in your parietal cortex are doing the work.

Scientists have measured photos of Einstein's brain and think his superlative parietal lobes allowed him to visualize his elegant mathematic proofs. Neuroanatomically, his parietal lobe was larger than those of normal healthy control brains. From a 1999 paper in *The Lancet*: "Visuospatial cognition, mathematical thought, and imagery of movement are strongly dependent on this region. Einstein's exceptional intellect in these cognitive domains and his self-described mode of scientific thinking may be related to the atypical anatomy in his inferior parietal lobules. Increased expansion of the inferior parietal region was also noted in other physicists and mathematicians. For example, for both the mathematician, Gauss, and the physicist, Siljeström, extensive development of the inferior parietal regions, including the supramarginal gyri, was noted."

Hidden within the center of the parietal lobe—in the deep groove that runs down the center of the brain, between its hemispheres—is the precuneus: Brodmann's area 7.

"Now we're realizing that the precuneus also expanded," says Coolidge, who, along with his coauthor, anthropologist Thomas Wynn, has written speculatively about prehistoric cognition in books like *How to Think Like a Neandertal*. Tucked away in an inaccessible spot, the precuneus is a difficult brain region to study. But with the advent of noninvasive imaging techniques, scientists are beginning to understand that a lot of its activity is related to space too. For instance, as we navigate an environment, we need to keep track of our movement through it—it's a process known as spatial updating. In a 2018 study, when researchers used transcranial magnetic stimulation—the same process Daniel Dilks uses at Emory University—to disrupt activity in the precuneus, they found that subjects could no longer spatially update the location of objects in their egocentric maps of space as they changed position.

In a 1998 case study, Japanese researchers described a seventy-year-old woman who had suffered a hemorrhage in the middle of her parietal lobe—specifically, in the precuneus. Waking up at midnight with a headache and nausea, she tried to walk to the bathroom of her house, but she couldn't reach it. She knew its location, but still it eluded her. Deep in her precuneus, a blood vessel was bleeding into the surrounding space. The blood, under pressure and with nowhere to go, was causing tissue damage. In tests, her allocentric reference frame was intact: she could draw an accurate map, describe a route, recognize familiar buildings, perform mental rotation tests, and learn spatial tasks. But her egocentric sense of space, which resides in her thoroughly modern and expanded human parietal lobe, was so damaged that she couldn't follow any of the maps she drew. She couldn't find her own bathroom.

If researchers asked her to describe the view of her house from a certain angle, she could do it—for instance, she could describe the view of the house from the backyard gate. If she was shown an image of the house from the backyard, an exercise that requires intact egocentric spatial reasoning, she couldn't tell interviewers the location the photo was taken from. She was unmoored in space.

•

In lots of other ways, anthropologists don't think we differed much from Neanderthals. The stereotype of the grunting caveman is long gone. "I don't see any need to deny Neanderthals speech or language," says paleopsychologist Coolidge. "I'd be really surprised if I went back—if I could go in a time machine, quietly—and they weren't asking each other questions."

Most anthropologists now agree that Neanderthals probably had some language abilities—if perhaps not quite as complex and rich as ours. They point to evidence such as the hyoid bone: a small and slender horseshoe-shaped bone that floats, suspended in the neck, and is a vital component of the anatomical apparatus required for speech. The Neanderthal hyoid bone was almost identical to the bone found in modern humans. Even hyoid bones from our common ancestor *Homo heidelbergensis*—such as a bone found in Spain, and estimated at 530,000 years old—indicate a capacity for speech. Researchers also point to the presence in our genome of a gene called FOXP2, which encodes a protein involved in language acquisition.

Neanderthals had the gene too.

Researchers expected to find that the DNA from Neanderthal bones contained an ancestral FOXP2 gene instead—more like the form of the gene found in chimpanzees. Not so. They found copies of the same gene as humans have. "The Neanderthals surprised us a bit there," lead researcher Johannes Krause told *Nature* when the study was published in 2007. But Coolidge thinks the language acquired by Neanderthals was subtly different from ours, lacking one particular form of speech.

"Here's what I don't think they had," he says, "and it's one of the highest pragmatics of speech: it's the subjunctive."

The subjunctive is a special verb form we use to express various kinds of unreality—counterfactual situations, dreams, wishes, myths, opinions, and future possibilities. We use it to describe actions that are provisional and events that might not even take place: the uncertain. If *that* happens, perhaps we could do *this*. That's the subjunctive. It was the form of speech that powered our symbolic thinking. If you bury your dead, their transition to the afterlife will be harmonious and trouble free. In other words, it's the language of gods and monsters. But Coolidge thinks Neanderthals didn't use it. He's convinced the physical brain changes that allowed for more sophisticated features of speech in humans—like linguistic recursion—also gave humans the ability to imagine and reason in ways Neanderthals couldn't.*

Why is this important? It begins to explain some of the potential differences between humans and Neanderthals. Archeologists point to evidence that, in some instances, Neanderthals buried their dead and suggest it provides proof of their subjunctive thinking. Coolidge disagrees. Neanderthals were more pragmatic, he says: they buried their dead because dead bodies smell. "I'm not sure they were thinking of the supernatural or creating all these fantastical gods that are going to take care of us in the afterlife like the Egyptians created," he says. "That's subjunctive thinking."

For hundreds of thousands of years, Neanderthals—pragmatic, rule-oriented, and subjunctive-less—had roamed their territory unchallenged. Then, suddenly, everything changed.

"These tall, gracile *Homo sapiens* come in: us, forty-five thousand years ago," says Coolidge, "with fully modern minds, highly ritualized burials, all kinds of subjunctive thinking."

Again, the subjunctive: the hypothetical. If the bison start to head off in a long unbroken line that curls toward the foot of the mountains, let's cut straight through the marshland and meet them there. "We made mental models in our parietal space, our visual space," says Coolidge, "with all kinds of verbal tags that allowed us better decision making, not just based on prior decisions—on the probability of success or

* When Coolidge talks about the subjunctive, it's a linguistic concept and not a grammatical one.

failure—but actually making up those models and then testing them in our minds. We used mental trial and error."

It made us unstoppable. Consider this: what do you get if you take a species like *Homo sapiens*, which processes the world and its contents symbolically, and then give it a rich, expressive language filled with subjunctive thinking with which to communicate? You get directions. Then, eventually, you get maps. Discovered in 1993, the oldest known map has been dated to around 14,000 years old. It was discovered at a dusty field site in Abauntz, near Pamplona in northern Spain.

•

Alysson Muotri (a 4 out of 10) has found another way to compare human and Neanderthal brains. At the University of California, San Diego, he opens an incubator and pulls out a culture plate. In a circular well, several pale little misshapen balls roll around gently in a warm broth of nutrients. The balls are brain organoids.

Six months earlier, Muotri had taken healthy human skin cells and deprogrammed them, converting them into stem cells. From there, the cells now had the potential to differentiate into any kind of human cell, but Muotri convinced them to become brain cells. In time, the cells began to differentiate and divide and organize, forming connections and networks, and became an organoid. Neurons met at synapses. In a sense, the organoids slowly became primitive little brains. After two months, Muotri can detect electrical activity. Eventually, the cells in each minibrain even begin to fire in synchronized bursts, like the spikes of activity seen in neurons in a functioning brain.

Each mature organoid contains around two million cells and is around the size of a pea. Muotri says he can keep them alive for several years. He uses the organoids to study conditions like autism and pediatric epilepsy. He's already seeing results: organoids begun with cells harvested from autistic children form different connections and networks than organoids grown using cells from non-autistic children. During the 2015 Brazilian Zika virus outbreak, researchers used organoids to show that the virus causes microcephaly in developing brains.

"One of my academic interests is human brain evolution—the idea that the modern human species is something quite unique," says Muotri. "We haven't seen it before. If you look at our closest relatives, the chimpanzees, they don't self-organize like we do, so the human brain is quite unique in that sense. We can imagine things. We create large societies. It's very hard to find anything similar in other animals. I decided that this is probably due to our high-level brain power and the question is: how did we achieve this high-level cognition that we don't see in other animals?"

Muotri makes the organoids act more like the brain of a Neanderthal—a process he calls *Neanderthalizing* the organoids. By studying DNA harvested from Neanderthal fossils, researchers have identified more than 200 genes that differ between modern humans and Neanderthals. Since we're so closely related, scientists reason that it's these specific genetic differences—some of them very subtle alterations in the genetic code—that ultimately make us human. Using cutting-edge CRISPR technology, a method of gene editing, Muotri was able to take one of those genetic differences and inserted it into the genome of the cells that form the organoid.

Small genetic changes can have large downstream effects. "There are many syndromes where you hit a single gene, or a single base pair, that creates a mutation," he says, "and then you have a human who doesn't talk, or cannot walk, or cannot learn."

Muotri began with a gene called NOVA1 that is involved in neural development. Neanderthals had a variant of the NOVA1 gene that we don't. It's no accident that the Neanderthal variant of the gene has disappeared, says Muotri. "As soon as someone acquired the modern version of the gene," he says, "the selection was so positive that everybody has that variant. We don't see it, the old version, these days—probably because there is no advantage in carrying the old version."

In other words, NOVA1 is so important for neural development that once the mutation occurred, we kept it forever. As they mature, the Neanderthalized organoids with the edited NOVA1 gene quickly begin to show their differences. They even look different—instead of the characteristic spherical shape, they're irregular in shape, like popcorn, says Muotri. "I was not expecting to see something so dramatic," he says, "but the morphology of these brain organoids was quite different, suggesting that the way the cells migrate to form the cortex, which is the region we decided to focus on, is quite different from what we have now."

Cells in the Neanderthal organoids make fewer synaptic connections with other cells. Even the electrical activity they produce is different. "The physiology of the cells is quite similar but the networks they form is very distinct, so you can tell right away, just by looking at the spike signals and how they evolve over time, that they do it in a different way," says Muotri. "I don't know if it's better or worse. It's just different."

From the complete Neanderthal genome, Muotri has selected another ten extinct gene variants he plans to edit into the organoids in the coming years.

•

With his *Sea Hero Quest* app, Hugo Spiers has shown that men and women can navigate equally well when they *are* equal. Even so, men and women employ different navigation strategies, says Mary Hegarty, a cognitive psychologist at the University of California, San Diego. These sex differences are so dependable that scientists have come up with an explanation for them, known as the *hunter-gatherer theory*, the origins of which go all the way back to our beginnings as a species—to the time that we became modern humans.

Hegarty has been studying sex differences in navigation for decades. In a 2018 study, she used a method known as the *dual-solution paradigm* to compare men and women. First, subjects learn to take a simple route through a virtual maze. Within the maze, landmarks are placed at different locations: a wheelbarrow, a duck, a chair, a piano, and so on. In the training phase, participants follow arrows that take them through the maze five times, past the landmarks without deviating from the training path. Duck . . . piano . . . telescope . . . chair. During the test, subjects begin at one of the landmarks (piano) and Hegarty asks them to navigate to another one (chair). Across many trials, a pattern emerged: men take shortcuts through the maze, but women stick to the routes they've learned.

Other studies have shown that, in addition to taking the shortcut, when compared with women, men pause less, travel farther without changing their course, and return to previously visited locations less often.

There are other differences too. Studies have shown that men tend to think about space in Euclidean terms, relying on metric distances, and cardinal directions like north and south—even if they're just trying to relocate their parking spot. Women navigate using landmarks instead.

For decades, one study after another has shown that men perform better at mental rotation tasks. In the lab, they're shown an image of a complex object made of little cubes and rendered in 3D. Then, they're shown an image of another object made of cubes and have to decide whether it's a new object, or a rotated view of the original object. Consistently, men outperform women at this test. But women are better at tests of object location memory: they remember where things are located. Whether the test is conducted in virtual reality or in the real world, as a group, women always eclipse men in cataloging the location of objects.

This is where the hunter-gatherer theory comes in. It goes like this: since evolution is an adaptive process, it selects for the traits that will help us and then reinforces them. Giraffes evolve longer and longer necks. Cheetahs evolve to run faster, and so do antelopes. In our prehistory, men—whose social worthiness depended on their hunting abilities—performed in ways that made mental rotation useful. For thousands of years, evolution reinforced that ability. But now, instead of using it to track mammoths through challenging and uneven terrain, men can mentally rotate a shape made of little cubes *in their minds* and determine what it would look like from different angles.

For a 2009 study, scientists even saw evidence for these differences at the neurobiological level. The cortical surface area of the parietal lobes—the brain region involved in tasks like mental rotation—is greater in men than in women. If we rewind tens of thousands of years to the Paleolithic era, women were likelier to be gatherers—not hunters. Mental rotations were less useful to them. Instead, women had to be able to recall the precise location of a fruit tree, or the place on a sheltered hillside where they could find the herbs that reduced a fever. For this reason, modern women are doomed to remember the location of school textbooks, and batteries, and homework folders, and library cards, and instruction manuals, and asthma inhalers, and passports, and spare house keys.

The problem with the hunter-gatherer theory is that it's complete nonsense, says anthropologist Ariane Burke. The differences are real enough—hundreds of different studies have reported the same patterns of ability. "Women in the Western world are much better at object location memory, and men in the Western world are much better at mental rotation," says Burke. "But I'm quite sure it's largely a function not of inherited sex-based differences in the brain, but rather gender differences in activity patterns."

In other words, these differences aren't the result of thousands of years of steady evolution, she says. They weren't determined by our ancestors. It doesn't take 40,000 years to undo them. In fact, it takes a few hours playing a video game to erase them. In a 2007 study, if women played an action video game for ten hours, the differences between them and men on the mental rotation test were narrowed. "The women kept improving until their skill level in mental rotation started to get basically as good as the men's," says Burke.*

Researchers have recorded the same effect in women who attended origami training sessions. Once a week, subjects were taught to fold paper into different complex shapes—a cube-shaped inflatable box;

* The hunter-gatherer hypothesis has held since the mid-1960s, but that doesn't make it right. There's plenty of accumulating evidence that women were capable hunters too. Archeologists have unearthed the remains of hunters wearing armor, alongside their buried hunting weapons, and, tantalizingly, in some cases, the hunters appear to have been female.

a crane with its wings frozen mid-flap; a complicated and intricate dragon that requires eighty-three separate folds in the paper. (Incidentally, I've studied the instructions to make the cube-shaped box—the simplest pattern the subjects learned. I might as well be studying the equations required to put a satellite into orbit.) After three months of training, the women's brains were changed by origami—dramatically altered, reorganized. Their mental rotation scores were now closer to the men's. They had closed the gap.

Origami had altered their brain activity too. When performing the mental rotation test, men usually show increased brain activity in the parietal cortex and the precuneus—areas that automatically process spatial information. In women, neural activity spikes instead in regions like the prefrontal cortex. In other words, it's more effortful, and costly, and inefficient. It's work. Once they'd perfected making a paper dragon, their brain activity had shifted, decreasing in the prefrontal cortex, and increasing in the parietal regions. Endlessly flexible, their brains had found a way to perform the test more efficiently.

In 2005, the BBC conducted an online survey to understand differences between the sexes. More than 250,000 people from fifty-three countries responded. The survey included a mental rotation test. Interestingly, in countries with the highest levels of gender equality—countries like Norway and Iceland and Sweden—the differences between men and women were the *largest*. These are the countries where Hugo Spiers, mining the mountains of data from his *Sea Hero Quest* app, saw the gap between men and women closing. Those findings are still true—just not true for mental rotation. Norwegian women performed mental rotations better than women from anywhere else in the world. They dominated when compared with women from countries like Pakistan, India, and Egypt. But Norwegian men still outperform them.

The latest research shows that while preschool boys show no advantage in mental rotation skills, they begin to overtake girls in the first years of formal schooling. Girls, and later women, never catch up. Equality doesn't necessarily lead to a sudden enthusiasm for video gaming or origami. Besides, these are statistical generalizations. While they tell us something useful about the differences in spatial abilities of men and women, they don't say anything about me as an individual. My wife's object rotation abilities far exceed mine, and so does her memory for the locations of objects.

As I stand in the darkness of the Human Evolution gallery at the Natural History Museum in London, I look again into the eyes of Spy II, the Belgian Neanderthal, and I wonder what he might have thought of a piece of paper folded eighty-three times, into the shape of a dragon. How would Spy II have reacted to a mirror maze?

•

We will never know why Neanderthals like Spy II disappeared. Perhaps we overran them or outcompeted them for vital resources. Maybe we were at war, or we just watched them slowly dwindle away to nothing. It's likelier that a combination of pressures all came at the same time.

But the expansion of the parietal lobe that we see in Ralph Holloway's endocasts could represent all the possible reasons Neanderthals might have struggled in our presence. It's why we formed long-distance trading relationships with other humans. It's probably the reason humans developed weapons like long-distance spears and bows and arrows—and the coordination to use them. Neanderthals didn't. When they disappeared, they were using the same stone tools and simple weapons they'd been using for hundreds of thousands of years. The cave art found at Paleolithic sites across Europe—the herd of broad-shouldered rhinoceroses that looks like it's about to stampede from the walls at Chauvet in southern France (estimated age: 32,000 years)—is called *parietal* art. In other words, art is an exaptation too. And so is a flute carved from a griffon vulture bone and punctuated with a neat row of perfectly tooled finger holes.

But our superior navigation skills are the exaptation that matters here. Suddenly, we had the ability to imagine ourselves into our memories, and our superstitions, songs, and myths. We could imagine ourselves farther along the track, or across space to the other side of the mountain—into possible futures.

Perhaps sudden mutations in genes like NOVA1 had a powerful effect too. Alysson Muotri will learn more via his Neanderthalized brain organoids—his Neanderoids.

Like a lot of other species—like Tolman's rats exploring a maze—we construct a cognitive map. Neanderthals did it too. They had the same place cells, grid cells, and head-direction cells as we do, and the same specialized cortical regions we use to perceive and interpret space. But, unlike Neanderthals, we can implant a cognitive map directly into another person's brain—via our symbolic imaginative thinking, and our expressive language.

My wife can construct a detailed inner map and then transfer it to me. It's a uniquely human trait. Perhaps Neanderthals even saw us do it. Maybe they watched us huddle together, heads close, pointing toward a mark on the landscape, a pale hill in the distance, and making shapes with our hands. They must have thought we were performing strange and powerful enchantments. Dark magic. We were conjuring something from nothing.

And perhaps we were.

Chapter Six

Dead Reckoning

The Sahara Desert ant (*Cataglyphis fortis*) doesn't like to take chances. It can't. If it gets lost in the relentless and extreme heat of the desert, even for just a few minutes, its body will overheat and literally cook on the sand. And so the desert ant has adapted to survive and navigate the otherworldly temperatures of its environment.

Even when the surface temperature of the desert is a broiling 178°F, the ant still ventures from its nest and searches for food. Less than a centimeter long, it is a foraging scavenger, going on regular hunting sorties across the sand to find the bodies of insects that have succumbed to the heat or died and were then blown by the wind across the desert floor. The ant works alone. It leaves the nest, traveling fast: fifty meters or so in less than a minute, hunting in widening zigzags across the sand; it pauses, turns, and travels another twenty-five meters in an imperfect arc; two minutes in the heat now, its body temperature rising dangerously into the red; a series of hairpin turns back and forth, searching; another minute spent foraging. Finally, hundreds of meters from the shade and safety of its nest, the ant finds a dead fly.

Far from home now, and in an inhospitable environment, the ant is like an astronaut adrift in space. Using its curving mandibles, it picks the dead fly up from the sand. The ant must find its way back to the nest—and quickly. It would be impossible to retrace its steps across the trackless desert. Even if it did, the journey would take too long in the punishing midday heat of the desert. It would die.

But the ant has been monitoring its journey across the sand the entire time—the widening spirals, the hairpins, the U-turns. Every step. It knows precisely where it is relative to the safety of the nest. Dead fly still gripped tightly in its mandibles, the ant travels to the nest in a straight line: the homeward vector. It takes the shortest path. It's an astonishing feat. Outbound: meandering and circuitous, like a drunkard on his way home. Inbound: as straight as an arrow in flight.

•

Meanwhile, in the forest reserve on Maui, Amanda Eller had tumbled down a steep rocky slope and fractured her leg. It was several days now since she had stepped off the trail and was swallowed up by the wilderness. The forest was full of noises. At night, the wind bent the tops of the trees and made an enormous sound, like an invisible ocean roaring above her. In the darkness, the temperature would suddenly drop and she would try to find places to sleep in the fog. She covered herself in leaves to stay warm. Eller drank from the streams that cascaded down the hillsides and into the ravines. She ate insects. But her odds of survival were slowly beginning to dwindle.

The morning after she disappeared, her boyfriend had reported her missing, and police scrambled, finding her car parked in the Hunter's Gate lot: the key hidden beneath a tire, water bottle on the passenger seat, purse with the cell phone in it placed in the passenger seat footwell.

Within a day, search parties had begun to fan out, dispersing through the forest, calling her name. Search dogs pulled their handlers through thick underbrush after a scent trail. A drone skimmed over the treetops. Helicopters. Infrared cameras. Eller's family offered a $10,000 reward for her safe return. In the forest, a volunteer killed a wild boar and checked its intestines for human remains, pulling apart the wet pearlescent coils like a fisherman untangling a knotted line. Perhaps a gleaming tooth, or a shred of Eller's yoga pants.

Eller's navigation abilities had failed her. Like a character from a Kafka novel, every time she tried to get closer to her car, she had walked farther from it instead. She had followed boar trails, walked into dead ends, and taken paths that led her nowhere. And then came the fall: deep in the forest, and farther than ever from the trail, Eller had tumbled awkwardly down a steep slope and fractured her leg. Things got worse still. It rained. Biblical rain. In the deluge, Eller lost her shoes.

On the fifth day, with no sign of Eller, the official search was scaled back. Maui-based journalist Breena Kerr reported on Eller's disappearance for the *New York Times* and the *Washington Post*. Months later, Kerr sends me smartphone videos of the trail. In places, the rust-colored trail is almost indistinguishable from the rest of the forest. I can easily imagine stepping away from the path and never finding it again. Even so, volunteers—now more than a hundred of them—continued to scour the forest, following streams and skirting ravines, peering into their depths to search for Eller. The effort became hi-tech. Back at their basecamp in the Makawao Forest, volunteers uploaded GPS data from their smartphones, displaying the precise search routes they had taken through the forest that day on a bank of flat-screen monitors.

But Amanda Eller, adrift and now badly injured, had managed to evade them all.

•

The desert ant finds its way back home across the salt pan by dead reckoning—a system of navigation that requires no external information, landmarks, or even a cognitive map. The signals all come from within. When the time comes to return home, the ant simply computes the shortest route back to its nest based on the path it has already taken. With no ability to reason or think, and only a natural preference for staying alive, this desert ant can perform a complex Pythagorean calculation from the data stored in its own body. It can return home in the dark if necessary. It can navigate an empty desert landscape completely devoid of landmarks.

The modern term for dead reckoning is *path integration*. In an 1873 letter to the science journal *Nature,* Charles Darwin was the first person to suggest that animals might use it to navigate. For centuries before that, *dead reckoning* was a term used by mariners who deftly integrated their rate of knots, time elapsed, and last known position to determine their current location at sea.

In order to understand the dead reckoning system the desert ants use, scientists have tested it every possible way—every permutation. The research published on desert ants could fill an entire library. Harald Wolf, a researcher at the University of Ulm in Germany, has sent ants across the salt pans wearing little eye caps to make sure they can't see. He and other researchers have captured ants and made them walk across slippery surfaces, or up inclines, or through complex mazes, or overburdened with heavy objects to carry.

Nothing stops the ant from calculating its home vector and traveling along it.

For another series of experiments, researchers constructed enormous surreal obstacle courses for the ants to tackle. Made of steel, and looking more like postmodern sculptures, the sunlight bounces off them in the desert near Maharés, Tunisia. One model, known as a linear array, resembles a strange horizontal staircase zigzagging across the wavering heat of the salt pan. It allows researchers to determine whether ants record uphill and downhill movement differently, or less accurately. They don't.

For yet another study, investigators built a steep ramp that exits directly from the ant nest on the ground, rising like a silver track into the sky, where it rounds a sharp bend supported by crisscrossed cables to an elevated feeding station. It looks like a roller coaster, but for ants. Researchers send ants trudging up its inclines to show that, even when they are climbing up and downhill, and through three-dimensional space, the ants still manage to calculate the ground distance they covered to get to the feeding station. They continue to update their home vector precisely. It's a monumental feat of computation.

Ant researcher Wolf says the ant is counting its steps. As it forages across the hot sand, it uses an internal pedometer to monitor the distance it travels—Wolf calls it a stride counter. If he carefully picks the ant up from the salt pan, a dead fly still gripped in its mandibles, and then transfers it twenty meters north and returns it to the sand, the ant will incorrectly follow its original home vector to a nest that isn't there. In other words, it travels in a straight line to a location exactly twenty meters north of the nest: the place where the nest *would* have been if the ant hadn't been moved. An experienced mariner would make the same mistake if his ship was suddenly lifted dripping from the ocean by Poseidon and dropped a thousand miles off-course in open seas.

The ant uses other information too. It builds a representation of the environment. As long as it's close enough to home, the ant uses olfaction to detect the elevated levels of carbon dioxide wafting from the enclosed confines of its nest. It monitors the direction of the wind currents that blow across the superheated desert surface. Specialized cells lining the uppermost rim of its compound eyes detect the direction of polarized light—a sensory system known as a sky compass. Together, the ant's constantly accumulating step count and the directional information of the sun's azimuth—its angle relative to the horizon—point it toward home.

Wolf has devised an ingenious approach to test his stride counter theory. A *Cataglyphis* ant nest is small. These are not the enormous sprawling wood ant nests of Europe and elsewhere—nests that can house millions of busy inhabitants like seething underground megacities, like an insect São Paulo, or an ant Shanghai. Instead, the typical desert ant nest consists of a single queen—the foundress queen—and the 300 or so worker ants she has raised to help support and protect her. It's a village, not a city. At a hot and dusty field site in northern Tunisia, Wolf traps a hundred or more ants from a nest, catching them by hand. They're fast, he says: the Usain Bolts of the insect world. In one second, they can run a meter. In the summer, Wolf spends three months at the field site, running low across the sand, decidedly un-Boltesque, behind the zigzagging ants, which he manages to catch by hand.

Asked to rate his spatial abilities, Wolf says no human from the first world even scores better than a 6; aborigines score an 8, and on a good day he is perhaps a 4. Back in the desert, selecting a single specimen from a tub, Wolf pushes a sturdy ant into a lump of modeling clay, back first, so that its legs are accessible. He shortens the legs of some of the ants, carefully snipping off the last leg joint with scissors. Other ants get leg extensions instead. Wolf attaches a little stilt to each leg—a bristle from a pig bristle brush. For hours at a time, Wolf bends over squirming little chestnut-colored ants embedded in modeling clay, gluing pig bristle stilts to legs that are bicycling constantly in the hot desert air.

On release, the ants that Wolf has fitted with stilts overshoot the nest. They still take the right number of steps to get home—they have counted them out perfectly, like little accountants. But their strides are longer now. They cover more ground with each step. The ants with shorter legs take the right number of strides, but since their stride length has been shortened, they don't arrive at the nest. They undershoot it instead.

Scientists have shown that other animals dead reckon too. Fiddler crabs and other crustaceans do it. The Namib Desert spider (*Leucorchestris arenicola*) constructs its cognitive map during daytime learning walks and then relies on path integration to find its retreat at night. Bees will follow visual landmarks, but when they fly above unfamiliar terrain—when they leave their map—they resort to another kind of dead reckoning, relying on optic flow data, literally a record of the information that passes in front of them to gauge distances.

Blind mole rats dead reckon in their complex burrow systems, which corkscrew and radiate underground in all directions like inverted subterranean cities. In the darkness of the soil, they're constantly integrating their internal signals—the record of their own movement—with information from the Earth's magnetic field. They're reading the planet: an Earth compass. Bats navigate via fixed routes known as flyways. These often consist of linear features, like darkened hedges, or the simple geometry of a country road that divides farmland like a long ribbon, or the sinuous curve of a river in moonlight. But a bat needs to know how long it followed a flyway before it stopped to forage for insects, leaving the flyway like an exit ramp from a highway: it uses dead reckoning. When they are first shown the location of a food target, dogs can even dead reckon while wearing a blindfold and earphones that play white noise. They do it by integrating information gathered during locomotion.

Humans dead reckon constantly, collecting and compiling information about our moving bodies—the idiothetic data. We can mine the information generated by our own bodies to tell us where we are. In his 1873 letter to *Nature*, Darwin mentioned an earlier account from the intrepid Russian explorer Ferdinand von Wrangell. A portrait of von Wrangell hangs in the Hermitage Museum collection: a balding man with a pinched face bookended by white muttonchops, pale watery eyes behind spectacles. He's wearing a blue military uniform decorated with thick gold braid and medals. It's an Imperial Russian Navy uniform—von Wrangell was an admiral. In 1820, he commanded the Kolymskaya expedition, sailing north through a white curtain to explore the Russian polar seas.

Darwin must have read von Wrangell's account, published in 1840, for in 1873 he wrote:

> With regard to the question of the means by which animals find their way home from a long distance, a striking account, in relation to man, will be found in the English translation of the Expedition to North Siberia, by Von Wrangell. He there describes the wonderful manner in which the natives kept a true course towards a particular spot, whilst passing for a long distance through hummocky ice, with incessant changes of direction, and with no guide in the heavens or on the frozen sea. He states (but I quote only from memory of many years standing) that he, an experienced surveyor, and using a compass, failed to do that which these savages easily effected. Yet no one will suppose that they possessed any special sense which is quite absent in us.

The landscape was an enormous blank sheet of paper. To von Wrangell, it was devoid of landmarks. No data. White on white. But the indigenous people in Russia's Arctic north, just like the desert ant, were never lost. They could read the blank paper like a map.

Darwin continued:

> We must bear in mind that neither a compass, nor the north star, nor any other such sign, suffices to guide a man to a particular spot through an intricate country, or through hummocky ice, when many deviations from a straight course are inevitable, unless the deviations are allowed for, or a sort of "dead reckoning" is kept. All men are able to do this in a greater or less degree, and the natives of Siberia apparently to a wonderful extent, though probably in an unconscious manner.

Unlike the Siberians, I am not able to do this to any degree at all. Von Wrangell would have watched me with dismay. I can navigate neither hummocky ice nor my own neighborhood, which bristles with familiar landmarks.

•

Elizabeth Chrastil (a 9 out of 10) is standing in an empty, low-ceilinged room on the campus of the University of California, Irvine in the Center for the Neurobiology of Learning and Memory. In fact, it's not even a room: it's a portable classroom trailer, a room that can move, not unlike the rooms of my own house when I try to imagine them in my mind. It sits approximately forty miles southeast of the concrete freeway sprawl of downtown Los Angeles. The skylights and windows have been blacked out, and the walls are unadorned and painted a nondescript grey color—approximately, Pantone Cool Gray 6C. The carpet is an institutional pistachio green. At the moment, in the center of the room, one of Chrastil's subjects pulls a sturdy black pair of virtual reality goggles over her head. Tucking in her disobedient hair, she straightens and then lifts her foot and takes a tentative step forward—into the void.

Chrastil is leading her subject through a virtual path integration test. Outside the trailer, beyond its thin walls, the campus bustles with activity as students head from lecture halls and laboratories to the science library and bagel shops. But inside the trailer, the subject is navigating a featureless pixelated desert.

"A lot of people are really excited to do virtual reality tasks," says Chrastil. But then they land in the virtual desert. It's not fun. It's like *Tron*. "It's actually really kind of boring," she says. Even so, to Chrastil the featurelessness of the virtual desert is an important tool: it's an empty environment. "We want to reduce the number of cues and landmark information that are out there in the world," says Chrastil. We rely heavily on landmarks when we navigate, and there are brain structures solely dedicated to detecting them. But landmarks won't help Chrastil understand path integration.

"They can be very useful for navigating," she says, "but they might not force you to rely on path integration."

This is not an abstract idea. It has concrete real-world utility. Imagine you're staying in an unfamiliar city: early in the morning, you leave your hotel and step onto the sidewalk, into the shadows of the canyon formed by two facing walls of skyscrapers. It's a place the sunlight never quite reaches, like underwater depths far beneath the surface chop. You are in search of coffee. You turn left, walk to an intersection that belongs to a gang of scruffy pigeons, and then take a right, where you pass a pharmacy, a shoe store, and a watch repairer, before turning left again, onto a wide avenue lined with brand-name stores. Nike. Apple. Gap. Tiffany & Co. And then finally: a coffee shop, sandwiched between dark and shuttered designer stores.

By this time, if you are like me, you are hopelessly lost. After zigzagging across the quiet city in the morning, I can no longer retrace my steps and return to the hotel. If I could drift high above the streets and view its geometries like a map, or like one of those drone-shot scenes in a documentary, with the shadows of trees falling across the roads, and a score by Philip Glass, perhaps then I could find my way back. But I am in its abyssal depths. Ground level. Like one of Tolman's rats, I'm in the labyrinth.

My wife experiences a morning coffee run in an unfamiliar city differently. She simply obeys a deep, primal, intuitive sense of location that propels her back through grey streets to the hotel, with a still-warm coffee in each hand. It comes unbidden from the primitive parts of her brain, an orchestrated signal formed between different brain structures, like a bass line that underpins a melody, squirming and bubbling beneath it. That part of my brain is silent. Flatline. Like a tumbleweed bouncing across a vacant lot in the dark.

How are we different?

Back in a trailer in Irvine, Chrastil is leading her subject carefully around the room. She steers her in a wide loop across the pistachio green carpet, their arms loosely interlinked, like an elderly couple on an evening stroll. Slowly, she walks the subject from one X, marked in pink tape on the floor, to the next.

"I usually define path integration as the continual updating of position and orientation in the environment as you move through it," says Chrastil. One of the most important functions of path integration is to keep track of a home position—the ant nest in the desert, or the hotel after a coffee run. But, says Chrastil, humans just aren't that skilled at it.

"We're not as good at it as a desert ant or some of these other species that are really great at path integration," she says. Generally, the human brain has evolved to navigate via other more sophisticated spatial systems. "In addition to our path integration systems, we really rely very heavily on these other systems," she says, "like landmark use, planning, goal-directed behavior, and a lot of other things that involve a lot of other parts of the brain that we have access to, and that have these higher order capacities."

Guiding her subject to another pink X on the floor, Chrastil is testing her ability to maintain a home vector using a navigation task that she devised. She calls it the loop test: "We take people in a circle and we ask them to basically tell us once they've returned back to their start location," she says. "One of the nice properties of a circle is that you go farther away but then you come back again." While some people are very good at this task, others overshoot or undershoot their starting positions just like Harald Wolf's manipulated ants—and sometimes by enormous margins. Via fMRI, Chrastil can show subjects a video version of the loop test too, monitoring changes in brain activity as the subject travels around a virtual loop in the scanner. There in the brain, Chrastil has found a homing signal. It's a strange and mysterious finding—like a flickering light in the darkness.

"Certain parts of the brain—particularly hippocampus, retrosplenial cortex, and parahippocampal cortex—are all sensitive to the distance you are from your home location," she says. "These parts of the brain have a signal that increases the farther away you are from home and then decreases again as you get closer." In other words, they respond not to how far we've traveled—our cumulative distance—but to our Euclidean distance from home. "These regions aren't tracking how far from home you travel," she says. "That's a different measure. What they're sensitive to is how far you are from home." It's an important distinction. The distance *traveled from* a place is not the same as the distance *from* a place.

These are the regions that either fail individually, or fail to communicate with one another in a larger network, when I leave a hotel in search of coffee and end up in a dead-end alley drinking cold coffee. Together, specialized neurons in these brain structures might be tracking distance, says Chrastil, or time, or something else entirely. Either way, trying to determine the relative contributions of each brain region is a challenge.

"We have a whole brain," says Chrastil. "We're using it all the time to do all kinds of stuff."

•

In the virtual desert, deprived of landmarks and other useful data to navigate by, Chrastil's subject arrives back at her starting point and closes the loop, but then continues, overshooting it.

"A big piece of my work is understanding why some people are really good at navigating," says Chrastil, "and why some people are so terrible at navigating—especially when it seems like a really important skill. Evolutionarily, it's pretty critical, but some people are really bad at doing it." With a series of studies, Chrastil had been untangling path integration to understand how the brain monitors movement. Visual information is important, she says. Just like bees, the human brain registers movement through space visually via optic flow, a measure of the amount of information that passes in front of an individual's eyes. It's simply a record of the amount of stuff that goes past us, like video footage captured from the window of a moving car. "An example is actually *Star Trek*," says Chrastil. "When you go into hyperspace, and things are going really fast past you and you get this vection feeling from the optic flow."

Scientists haven't yet determined whether blind people, who are deprived of the information provided by optic flow, can navigate by dead reckoning as well as sighted people. Nevertheless, researchers *have* shown that blind people construct detailed and informative cognitive maps and can use them to navigate.

Body movement information is important too, says Chrastil—the idiothetic information. "We also have body-based information, from our legs primarily, but also from our vestibular system in our heads." These two sources of information are equally important in tracking our movement through space, she says.

And there are other kinds of information too, says Thomas Wolbers (8 out of 10), a researcher at the German Center for Neurodegenerative Diseases in Magdeburg, Germany, who studies path integration. "It's a multisensory experience," he says. The environment around us is brimming with landmarks and other spatial data and we gather them all—even, says Wolbers, tactile information like the textures on the ground, and the speed at which they pass; and the auditory cues we hear as we move through a space. These are the information sources that blind people use to build their internal maps.

Just how we integrate all of this spatial information and combine it to compute our location remains unknown. But it's how we maintain our position in an unfamiliar city, or a grocery store, or as we navigate a virtual desert in a trailer on the UC Irvine campus. And it's vital for learning the layout of a new environment, says Wolbers. "When it comes to learning novel spaces," he says, "being able to relate different locations in a novel environment to each other most likely involves a lot of path integration."

Chrastil uses the loop test. Other researchers use the triangle completion task. It's simple enough: in a virtual environment, subjects are asked to walk along two sides of a triangle and then turn to point toward their starting location. When they do this, several specific brain regions burst into activity. They are the usual suspects, says Chrastil: the hippocampus, the retrosplenial cortex, and the parahippocampal cortex. A subject whose right hippocampus is more activated during the triangle completion task can walk two sides of a triangle, turn, and point with greater accuracy toward her starting point than a subject with less activation in that brain region. Chrastil has shown that the actual volume of the brain regions involved is greater in people who are better at path integration.

One fact is tantalizing, mostly because it appears in scientific reports almost as an afterthought, buried within the tables, bar charts, and graphs: in any group of test subjects, some individuals perform so badly at the triangle completion task that they're thrown out of the study. The outliers, the ones kicked out, walk carefully along two sides of the triangle and then turn and point in completely the wrong direction. On the loop test, they're hopeless. They perform so badly that, just by including them in the data, they skew the numbers and obscure important findings—so they're removed. They are the noise that overwhelms the signal.

I already know that I am one of those people, forever pointing in the wrong direction. I am the noise. The static hiss. The mind like a clenched fist. I am an outlier, skewing the data.

Was Amanda Eller an outlier, too?

•

The human brain—or any brain for that matter—is not a homogeneous thing. It's divided into grey matter and white matter, which are found in approximately equal portions in humans. Grey matter consists mostly of neurons: for instance, the cell bodies of place cells encoding location in the hippocampus, and grid cells and head-direction cells in the entorhinal cortex, but also the ninety billion or so other neurons found throughout the brain. But white matter is not cell bodies. Instead, it consists of tracts of long-range nerve fibers—like bundled telecommunication cables--that connect a neuron in one brain structure to the branching dendrites of a neuron in another part of the brain.

Imagine that each neuron in the brain is a house. Like any house, each cell is complex, specialized, often defined by its location and its architecture, and influenced by its immediate neighbors. A cluster of neurons might represent a specialized brain structure like the hippocampus, in the same way that a subdivision of houses, while made up of individual structures, becomes its own thing. In this analogy, the white matter is the highway system that connects the houses and subdivisions to one another. White matter projects and swerves through grey matter. Like cars traveling between neighborhoods on a busy interstate, nerve impulses hurtle along it. Path integration is an orchestrated event. Impulses ping back and forth on high-speed tracts between the medial prefrontal cortex, the hippocampus, the retrosplenial cortex, and a region scattered across the surface of the cortex, known as the human motion complex.

When scientists use an fMRI scanner, they're only measuring the brain activity in grey matter, in the neurons—and grey matter only represents around half the brain. It's like studying the complex dynamics of a vast, sprawling, interconnected city by watching the lights turn on and off in the windows of a few individual houses. Chrastil wondered: what if, instead of focusing on scattered houses in distant neighborhoods, she studied the patterns of the traffic hurtling between them? In other words, what if she looked at the white matter?

In an ongoing study, Chrastil has begun to measure what happens in white matter when people try to path integrate, using a technique known as *diffusion weighted imaging*. "It looks at how water is diffusing in the brain," she says. "If it's diffusing evenly, that means nothing is constraining it." When water molecules move freely across brain structures in all directions, it means there are no barriers obstructing them. But that's not always the case. Sometimes, water molecules are blocked by obstacles in the brain—and this gives Chrastil important clues about the connectivity between different brain regions.

"If water is diffusing in a particular direction," says Chrastil, "it means it's on a white matter tract." In other words, it's on a highway. It's traveling along one of the high-speed corridors that swerves through the brain. Using this method, she can begin to map out subtle individual differences in the white matter of her subjects. "We can look at white matter differences between people who do well and people who do poorly," says Chrastil.

Perhaps the differences between my wife and me lie in the subtleties of our white matter, as it branches and insinuates itself between brain structures that function together as a unit—firing together during complex tasks like path integration. An otherwise well-designed city can be brought to a standstill by the small flaws in its traffic system. Gridlock.

One thing is clear: we don't all navigate in the same way. Chrastil thinks the strategies we each use might be determined by the differences in our white matter. People with higher levels of connectivity between their hippocampus and prefrontal cortex might rely on a different navigation strategy than those with a better connected retrosplenial cortex and occipital place area, Chrastil says. One person might rely heavily on landmarks, and another is much more successful at navigating via path integration, or processing verbal directions, or imagining their route through an environment using the same relatively modern parietal brain regions that separate us from Neanderthals.

•

Even so, dead reckoning has its limits, especially in humans. For us, it's an error-prone measure that should only be used in the short term. In the beginning, the errors are imperceptible, but eventually they begin to accumulate. As Amanda Eller discovered, the results can be disastrous.

In a 2009 study, Jan Souman, then a psychologist at the Max Planck Institute for Biological Cybernetics in Germany, tested the ability of people to walk in a straight line through unfamiliar terrain. *Walk a straight line*: it's one of the most fundamentally simple spatial tasks you can imagine. It recruits many of the same brain structures activated by path integration.

First, Souman took six subjects to the Bienwald, a region of unbroken forest in the Pfalz region of southern Germany. On a satellite image the Bienwald is a wide, spinach-green crescent of woodland surrounded on all sides by a tidy, peaceful checkerboard of farmland. It sits on Germany's rural border with France, a land of quiet green valleys and rolling hills. Subjects were taken into the middle of the forest—into the quiet shade of the trees—where they were fitted with GPS devices and instructed to start walking in a straight line. On a clear day, they maintained a straight line for up to four hours, skirting numerous obstacles, clambering over felled tree trunks, bisecting stands of centuries-old beech trees, but never deviating from their course. One subject walked so far that eventually he left the forest altogether. With the trees of the Bienwald rising behind him like a green wall, he began to walk out across the neighboring fields. He was on the checkerboard. When Souman turned the man around, he walked back into the Bienwald for several more miles—all of them in a straight line.

But on an overcast day, the results were very different. Without the sun for guidance, the subjects could no longer maintain a straight line. Chaos reigned. Their paths became untidy looping scribbles through the trees—as inscrutable as a doctor's signature. Viewed from above, the paths plotted via their GPS devices were lengths of knotted string thrown on a satellite image, half-unraveled. They double-backed on themselves. They snaked and looped. They walked in circles. Sometimes, without realizing it, they crossed their own paths—and then did it again and again. At times, the circles they walked in were as little as twenty meters in diameter. They knew neither where they had been nor where they were going. After four hours of walking through the quiet shadows of an overcast forest, most of them never traveled more than a mile from their starting point.

With the sun overhead, Souman's subjects maintained a straight line as well as any foraging desert ant returning to the nest. In fact, says Souman, we use the sun's azimuth just like they do. The sun is a compass pointing the way.

Souman repeated the experiment—this time in southern Tunisia, on the northern borders of the Sahara Desert. The same thing happened. There were no landmarks to navigate by, no roads, no coastline or distant peaks by which to orient—just miles and miles of undulating sand. It was the real-world version of Chrastil's virtual reality program. In daylight, the subjects managed to keep a straight course across the sand. Their paths were Euclidean in their undeviating straightness. They maintained a straight path at night too, in the pale glow of the moon on the endless sand—for a while, at least. But then the clouds gathered, moving across the face of the moon like an impenetrable screen. In sudden darkness, the subjects again began to scribble a random, aimless track across the desert, walking in circles and doubling back on themselves. It was like being in the Bienwald under grey skies.

Most of us can maintain a straight course, even across unfamiliar terrain, provided we have celestial bodies like the sun or the moon to navigate by. In this, we're not much different from—or perhaps even inferior to—the African dung beetle (*Scarabaeus satyrus*). A squat, black, round-bodied insect, the African dung beetle navigates using the same expertise, and seems bound by the same limitations too. When a large herbivore drops a load of fresh dung on the savannah, the assembled beetles scramble into immediate action. They roll the dung into large balls—often considerably larger than themselves—to eat and to use as shelter. The beetle is nocturnal. As a fresh plume of steam rises into the night air, the dung pile quickly becomes a battleground. It's gladiatorial. There is conflict and competition. The beetle must obtain its share of the dung but avoid skirmishes with competitors. It must claim its dung and escape.

The beetle has a very specific operating system. Once it's made its dungball, it clambers on top of it and carefully surveys the night sky. Then it begins rolling the ball across the savannah, away from competition—far from the maddening crowd. It does this by traveling in a straight line from the dung heap as quickly as possible. While desert ants return to the nest in a straight line via the home vector, the dung beetle does the opposite. It's outbound. It doesn't even care where it's going, as long as it travels away from the gladiatorial conflict of the dung pile. And the best way to do that is directly, in a straight line.

The beetle navigates by the polarized light of the Milky Way—the galaxy to which we belong. On clear nights, it appears as a diffuse, undulating band of light that unfurls across the sky. Pushing the dungball backward with its hind legs and facing the dung pile it's leaving, the beetle continually adjusts its route based on the position of celestial bodies. Roll the dungball. Keep it straight. Look at the Milky Way and the brightest twinkling stars. Adjust the dungball. Roll it. Look up again at the pale band in the sky that points the way like a road on a map.

Without the sun or the moon, the human brain will set its owner wandering. The same is true for dung beetles. For a 2013 study, researchers made little cardboard screens and glued them to the beetle's thorax so that the screen covered the uppermost part of its compound eyes. When a beetle can no longer see the sky—the glimmering stars, the Milky Way—its navigational system suddenly falls apart. The beetle begins walking a circuitous, winding path instead, sometimes going in circles, sometimes taking great time-wasting spiraling detours away from the dung pile. All of its dung-protecting strategies are erased. Scaled up to represent miles instead of inches and hours instead of minutes, their tracks look just like those made by Souman's subjects wandering the Bienwald on an overcast day. This is the same finding Elizabeth Chrastil made on subjects walking from one pink X to another, across a pistachio carpet on the UC Irvine campus: visual information and body-based information make an equal and independent contribution to path integration. They're both needed. By itself, one of these inputs isn't enough.

"Scale is really important," says Chrastil when I mention Souman's walking study. "It certainly suggests that our body-based information is not enough over really large distances. Body-based information might be sufficient for a room, but when you get into the scale of miles it really isn't enough." Instead, she says, we need visual information in order to keep realigning. "Our path integration systems are coarse and work okay for small distances, but they start to fade over time."

On a cloudy day, in a confusing environment like a dark forest, Souman's subjects had to rely on body-based information alone. It wasn't enough. His subjects were sent spiraling through the woods, crashing through green tangles of ivy as if each of them was a balloon that had been filled with air and then released to zip madly through space.

Chapter Seven

Somewhere East of Timbuktu

When J.N. finally decided to visit a psychologist, her husband jotted down a description of some of the fifty-six-year-old speech and language pathologist's spatial behaviors: *When at the movie theater,* he wrote, *if she goes to the bathroom, she does not know how to get back to her seat. Or, if she leaves a building and then tries to reenter it, she will not remember where the entrance is and has to walk around the periphery until she finds the entrance. She gets lost easily, and often and for years, [she] was accused of not paying attention.*

That sounds like me—or, at least, a slightly more severe version of me.

•

No one has studied topographical disorientation more closely than University of Calgary cognitive neuroscientist Giuseppe Iaria. For more than a decade, he's been recruiting patients like J.N., who are incapable of forming cognitive maps.

"About ten years ago," he tells me, "we described a developmental disorder that basically consists of people getting lost on a daily basis.

They get disoriented even within their own houses, from childhood, in the absence of any neurological condition or any brain damage or any congenital brain malformation, and in the absence of any other cognitive complaint."

In 2009, he published the first in a series of case studies that describe individuals with topographical disorientation. He called his first subject Patient One. (Patient One and J.N. are not the same person, and J.N. is not one of Iaria's patients. Scientists often anonymize subjects—think of memoryless H.M.—so as not to identify them.)

By the time Iaria met her, an inability to find her way had already shaped Patient One's entire life. As a child, she would panic in grocery stores the moment her mother left her sight, disappearing from view at the end of an aisle. Her sisters and friends always helped her to find her way. She never travels alone. Even now, as a forty-three-year-old, she lives with her father. She can recognize landmarks, such as distinctive buildings, but often gets lost in her own neighborhood and has to phone her father so that he can come and find her. He'll ask her what street she's on, and then use his fully functional cognitive map to rescue her from her maplessness.

Patient One's commute to work is a daily horror. Every morning, she leaves her house and takes the same well-worn route downtown. First, a bus into the city. The world scrolls past in the bus window and she watches intently, waiting for the moment the bus approaches a recognizable city square—a wide apron of concrete, with busy commuters rushing across it. Umbrellas. Taxis. City pigeons. The city is monochrome, a study in grey. But the square is an unmistakable landmark. When it slides into view Patient One knows to get off the bus and walk a short distance to a distinctive tall building, in which her office is located.

She cannot deviate from her route. If she does, she might get lost. A moment of inattention—noticing, for instance, that the trees next to the bus stop are all sporting little green buds instead of looking for the square—and she will be lost in the city, borne on a rushing tide of spatial anxiety as she tries desperately to locate the building she works in. But her employers are planning to relocate soon to a new building in another part of the city. After years of practice and repetition, her well-worn path will no longer take her to work. Now, finally, panic at the looming prospect of permanent lostness has sent Patient One to Iaria's lab.

•

Throughout history, doctors have described patients with symptoms similar to Patient One's, but their condition has almost always been acquired through injury, disease, or an unpredictable *event*. In an example from 1982, doctors described a seventy-two-year-old architect from Massachusetts with a brain lesion in his parieto-occipital region. His event was pathological growth. When the doctors asked the subject his location, he always told them he was in MGH, or Massachusetts General Hospital. He knew what building he was in. But every day the location of the hospital itself magically changed.

> Day 2: "My assumption is that it relates to the huge hospital, outside of London [England] or the London suburbs."
>
> Day 3: "It's a far extension of MGH in California."
>
> Day 4: "I know what you tell me but it seems to me I'm in Paris."
>
> Day 7: "It is a luxury hotel, somewhere in the Far East, probably Tokyo."
>
> Day 8: "It's a hotel in China or . . . it could be Japan."
>
> Day 16: "MGH East. . . . When I say east I'm thinking more of Baghdad rather than east Boston."

At other times he said he was in Arizona, or Denver, or in Africa. One morning, he was somewhere slightly east of Timbuktu.

Doctors and psychologists love people like the lost architect because his symptoms were simple to diagnose neurobiologically—and they were informative too. They told us something about how the brain works. H.M.'s hippocampus and the neighboring parts of his temporal lobes were surgically removed—vacuumed out in 1953. Is it any surprise he couldn't navigate afterward? When musicologist Clive Wearing's brain was ravaged by a viral infection in 1985, he was left completely unable to find his way, even in familiar environments. But his condition wasn't a mystery. Even less profound injuries can have similar effects. In recent work, Iaria has shown that young hockey players—aged sixteen or younger—who have suffered a concussion on the ice are less able to form cognitive maps than their non-concussed teammates.

Like H.M. and Clive Wearing, Patient One is superlatively lost. But hers is a lostness without a known cause. No injury. No hockey-related concussion. No infection damaging her hippocampus from within. Iaria began to wonder: Is topographical disorientation always an acquired condition? Is it always the result of a traumatic brain injury, or a burst capillary in a brain region that decodes space? Does it only occur when a tumor slowly expands, one cell at a time, into a finite and enclosed brain structure?

Consider a similar problem in a different cognitive domain: prosopagnosia, face blindness. Sometimes, like the lost architect's inability to navigate, it's caused by a localized injury or a brain bleed. But occasionally, the problem is congenital: people are just born with it. Perhaps, Iaria reasoned, the same is true for a lack of spatial abilities. Maybe some people with a profound and perpetual lostness—like Patient One—are just born that way.

These are the people he's interested in. And, over the last decade, he has found them. "They have a job, and they're functional, except that they have a very selective problem finding their way around, and they get lost in extremely familiar surroundings from childhood," he says. "There was nothing in the literature about developmental disorders of this kind, so we termed this condition *developmental topographical disorientation.*" Otherwise known as DTD.

From his research, Iaria estimates that perhaps as many as 1 or 2 percent of us might have DTD—or about the same percentage of people with prosopagnosia, he says. They are the people who rate themselves a 1 or 2 out of 10 when they're asked to assess their spatial skills. Iaria says he got into the field of cognitive neuroscience because he was interested in the incredible amount of variation in human brains. What drives it? What makes one person so much better at a particular cognitive task than another person?

He decided to study spatial abilities because of the opportunities it provided to look at person-to-person variation. Ironically, in the process he became the de facto world expert in DTD, which really doesn't explain normal human brain function at all: it describes the complete failure of the one to two percent of us who simply cannot navigate—the people clustered at one end of the spectrum, but not the spectrum.

"The majority of people who have this problem have a complete inability to form mental maps of very, very simple environments," Iaria tells me. "It doesn't matter how long they've lived in their house, or their neighborhood, or in the building where they work. It can be twenty years. Fifty years. There is no way they can represent in their mind where things are around them, and so even if they have lived there for a long time they always need to look around for clues and signs." They rely on clues and signs. Portents. Shadows. Vibrations. Hunches. Urges. Impulses. Fear. They rely on chance, and often they're wrong. Most frequently of all, though, they rely on memory.

"These people—most of them, the majority of them—have no problem learning how to go from A to B by remembering a sequence of paths and turns: turn left, turn right," says Iaria. "They have no problem recognizing landmarks. They have no problem making associations between landmarks and body turns, like turn right at the bank, turn left at the bakery, and so on. Despite that, they do get lost in familiar surroundings on a daily basis. So: why is that?"

Finally, in the last decade, we have begun to understand why.

•

In 2015, when J.N. began to look for help with her lifelong condition, she ended up in front of a team of researchers, at Carnegie Mellon University in Pittsburgh. There, Elissa Aminoff, a cognitive neuroscientist, decided to take a closer look at the functional and structural activity of J.N.'s brain. It is one of the most complete and thorough studies yet of a person with topographical disorientation of an unknown origin. Aminoff decided to focus on the brain regions that Daniel Dilks at Emory University had called Sceneworld. "With the study of J.N., what we really wanted to do was pull apart different aspects of navigation and see what she had issues with, and what we knew about the brain areas," says Aminoff, now a researcher at Fordham University in the Bronx.

The obvious places to start looking are the various components of Sceneworld: the parahippocampal place area (PPA), the occipital place area (Aminoff calls it the transverse occipital sulcus), and the retrosplenial cortex. These are the parts of the cortex that help us to decode the spaces around us. The first tests that Aminoff ran on J.N. determined whether her cortical scene-processing regions were working the way they should, allowing her to perceive her environment normally.

"We wanted to try to use our ideas about the functional roles of these three regions and see how that manifested in the behavioral performance of J.N.," she says. "Is this an issue of being able to identify what's in her environment? That's the first thing to rule out."

J.N. could see and recognize landmarks. She could differentiate a kitchen or a bathroom from a beach or a field. She understood that a door was categorically different from a painting. "It became very clear to us that there was not an issue with perception," says Aminoff. "It was a more advanced issue of being able to use that information, and manipulate that information, and abstract that information."

Gradually, Aminoff began to focus on J.N.'s retrosplenial cortex. "How I think about the retrosplenial cortex," she says, "is that it's a place that aggregates information. It takes information that the PPA might be perceiving as different viewpoints of the scene and puts it all together to understand the greater environment."

It's a brain region that makes order from disorder. It distills and edits information and makes it meaningful. And it's completely vital.

It doesn't matter if you're an early human radiating outward from Africa into unknown places, or a modern human relying on your procedural memory to visit a familiar place like the bathroom in your house. You need a functioning retrosplenial cortex. The man who tried repeatedly and failed to find the bathroom in his own home, for example, had suffered a ruptured blood vessel on the surface of his retrospenial cortex. Aminoff ran what's known as an adaptation task on J.N.'s retrosplenial cortex.

"There's this neural phenomenon that if you see something a second time your brain is not going to activate as much," she says. "It's called adaptation. When you see this adaptation profile—this lowered activation—it's a sign that the region cares about this information."

For instance, the PPA is hardwired to respond to places, but not to faces. If you show someone in a scanner an image of a location, you can see that the PPA responds strongly the first time it sees the image but less strongly the next time. Since it perceived and interpreted the spatial information in the image the first time, it doesn't need to repeat that same process every time. The brain maintains efficiency by not performing tasks it's already completed once. But this isn't true when you show an image of something the PPA is not hardwired to respond to—like a face. In that instance, there's no adaptation to the image.

J.N.'s retrosplenial cortex was abnormal, says Aminoff. "She didn't have any adaptation in the RSC for scenes," she says. Control subjects did. J.N.'s other scene-processing regions did adapt to the categories of objects that they're hardwired to perceive. We rely so heavily on these brain regions to help us understand the world that, when they fail, the effects can derail our ability to navigate even a simple environment, for example by following a simple map through a large city.

Next, Aminoff ran what's known as a resting state scan on J.N.

"One of the ways that you can look at functional connectivity—and look at whether brain regions are actually speaking to each other—is to take a scan when participants aren't doing anything," she says. Thus: the resting state scan. Subjects in a scanner are shown a blue screen with a cross in the middle. They're told to stare at the cross and try not to think of anything. Disappear. Empty the mind. Become the tundra. What happens when the brain isn't engaged?

"By and large, everybody has resting state networks where these regions are talking to each other," says Aminoff.

But not J.N. Her retrosplenial cortex seems almost functionally disconnected from her other scene-processing regions. The structural connections exist. The white matter is there, says Aminoff. But the brain regions don't communicate with one another. Imagine a building crisscrossed with phone lines, but some of the connections don't seem to work properly.

In his brain scans, Iaria has made similar observations. "What we have found is that different regions—the hippocampus, the prefrontal cortex, the retrosplenial cortex, the posterior cingulate, the posterior parietal—those regions are functioning okay," he says. "But when they need to get involved in forming cognitive maps, the communication between them is not as good as it is in the group of controls who have no complaints about orientation problems."

The brain structures that construct a cognitive map are delivering monologues instead of having a conversation with one another. Since only single case studies have been published, says Iaria, not much more can be said about the specific causes of DTD. On some deep level, even as a child, J.N. has always been aware of her deficit.

J.N. is Janice Nathan, a speech pathologist who lives in Pittsburgh and specializes in autism. In a self-assessment, she rates her spatial skills between a 1 or 2 out of 10. One afternoon, I sit in my sun-spangled office and talk on the phone with Nathan about our spatial shortcomings.

"I was always getting lost," she says. When Nathan was twelve years old, she got lost while hiking a remote Hawaiian trail. Coincidentally, it happened on Oahu, in the same dense tropical wilderness that swallowed up Amanda Eller in 2019. Nathan was a Navy brat, and at the time, her father was stationed in Honolulu.

"We belonged to a hiking club when I was a little girl," she says. While hiking with her father and brother, Nathan walked ahead along the trail, angry at her brother for teasing her. On her own, she arrived at a fork in the path. “The marker had fallen off the tree or whatever, so, of course, I went the wrong way,” she says. Suddenly, Nathan, her brain misfiring, was lost in sugarcane fields.

“I couldn’t backtrack,” she says. Instead, she forged on. She panicked, running haphazardly through sugarcane fields for what she now thinks was six or seven hours. Her family reported her missing. A search was initiated. Dogs. Helicopters. Finally, at night, the unmistakable sound of a highway drifted toward Nathan through the sugarcane, along with the reassuring headlights of passing cars. Roads mean one thing: civilization. On seeing a disheveled tween in their headlights, a family stopped their car, picked her up, drove her home, fed her, and called the police.

“Those kinds of things only seemed to be happening to me,” she says. “I really thought that was it. I was never going to be found again.”

As an adult, the struggle has continued. In 2001, Nathan moved to Pittsburgh, a city that she calls a navigational nightmare. She has lost purses. Lost herself. Driven past her own house. She has spilled many drinks, often in her own lap. It’s worth mentioning the spilled drinks in more depth. Research published in 2020 by a team at Johns Hopkins University shows that our grasp of intuitive physics—in other words, how things roll, swing, bounce, balance, slosh, slide, and collide—is strongly predicted by the strength of our spatial skills. Not everyone has the same internal sense of how the physical world works—of how not to slosh when you don’t want to slosh.

Nathan’s father, she says, had good spatial abilities but evidently didn’t pass them on to her. “He didn’t understand why I wouldn’t make my mental map,” she says. “He was always telling me: *use your mental map!* I didn’t even know what that meant. What’s a mental map? I would love to have one.”

•

If people with topographical disorientation like J.N. and Patient One can remember and can follow a sequence of instructions, why is that not sufficient? Why is *turn right at the bank, turn left at the bakery, and so on* not good enough for daily life?

"It would be too much of a load on our memory," says Iaria. "How many pathways do we need to remember? Even for the same pathway, we'd need to form two different sequences, because coming back and going there are two different ways."

This is why, beginning around the age of nine for most of us, we construct and rely on a cognitive map instead. We build it using landmarks. In one study, Iaria performed resting state fMRI scans (*empty the mind*) on DTD subjects and showed a decreased functional connectivity between the hippocampus and the prefrontal cortex. This, he says, affects an individual's ability to monitor and process spatial information while moving through an environment—to build the map. When we navigate within an environment, we need to simultaneously recognize its landmarks, understand the perspective from which we're seeing them, and retain the position of our target location.

This approach is dynamic, adaptable, flexible, effortless. In other words, it's everything that memory is not. "It's very, very quick," says Iaria. "But it's possible only because people have a mental map of where things are."

After first describing Patient One, Iaria went on to recruit 120 subjects with DTD for another study, testing their navigation skills and cognitive abilities. He began to carefully describe the condition.

One fact stuck with Iaria: a third of his DTD subjects reported that one or more first-degree family members also suffered from the same kinds of spatial problems they did. That's a lot of lost people, who are related to a lot of other lost people. It made Iaria ask whether genes might be involved in DTD.

He recruited entire families and began to look for patterns in the presence of DTD. Starting with a DTD-positive *proband*—the term for the person who is the starting point of a genetic study—Iaria drew complex multigenerational family trees, testing parents and siblings and children on their spatial abilities. It quickly became clear that DTD aggregates in some families but not others. It clusters.

Take Family 32: Iaria's proband in the family was a woman with DTD; she has two children, one of whom also has DTD. The proband's first cousin has it too. And so does the cousin's mother—the proband's aunt. In other words, members of at least three different generations in one family are unable to form a cognitive map. The proband's parents might also have it, but her mother's amnesia and a long-ago head injury suffered by her father make diagnosis impossible. But the thread of lostness runs through the family, from one generation to the next. And it runs through other families too.

After studying several of the families, Iaria noticed something else: every DTD-positive subject had at least one parent with DTD. The perpetually lost apple doesn't fall far from the tree.

Iaria then decided to take a closer look at the non-DTD relatives of his DTD subjects. Since they didn't report being lost in familiar places—an important diagnostic criterion for DTD—they were considered normal. But they really weren't.

Around half of them failed Iaria's Cognitive Map Formation Test. For that test, they were given twenty trials to form a mental map of a virtual environment. After several trials, most people begin to make a cognitive map of the environment. They recognize distinct buildings as landmarks and construct the map around them. On average, Iaria's control subjects finished the task after around nine trials. When the non-DTD offspring and siblings of DTD subjects tried it, half of them failed—even though they were considered normal. By their twentieth trial, they were still as inept as when they started the test. Maybe, eventually, they would have formed an inner map—on the thirtieth trial, or the one-hundredth, or one-thousandth. Perhaps not. They have developed strategies that allow them to navigate familiar environments—to approach normal. They have learned to cope. But they're impacted by the thread of DTD that runs through their family too.

•

Around this time, taking a break on social media one day, I found a Facebook group called "Directional Disorientation (aka Developmental Topographical Disorientation)." On the group page, members gather to share stories, ask advice, and vent frustrations. Until that moment, I had felt quite alone in my shortcomings. I had spoken with J.N. and felt the comfort of our shared experience, but had never known anyone who struggles to navigate in the same ways I do.

Now I know others. We are legion.

One day, someone posts:

> You have no idea (well, you probably do) how many wrong turns I took today, even in familiar places.
>
> *Another member posts:* When I follow my GPS and it tells me to turn now—but the map seems farther away still than the road in front of me—so I miss the turn. Am I the only one who does this?!?
>
> *Another member posts*: Hi all—random question but does anyone else get lost putting on duvet covers? No, I'm sadly not even joking . . . I have massive dramas with trying to fit a double duvet cover onto a duvet . . . No matter how hard I concentrate I almost always end up confusing corners and having to retry multiple times as I can't work out spatially how it fits. It's literally been an issue ever since I've had to do it—ahh feel much happier for sharing!

It's as unnerving as it is comforting for someone you've never met to describe your experience so accurately. I had found my tribe.

The Facebook group has more than 800 members. It was begun in 2013 by Andy Pakula, a minister at what he calls a nonreligious church in North London. Self-diagnosed, Pakula says he started the group so that he wouldn't have to feel so alone. Through the group, I meet people like Scott Kelbell (a 3 out of 10), an information software engineer from Wisconsin who tells me he has no sense of place. When he was fourteen, Kelbell was knocked off his bike by a car and suffered a concussion, and he has struggled with spatial tasks since then.

For months, Kelbell and I email each other as he designs a haptic compass—a wearable device that will buzz and vibrate whenever he faces north. By wearing his device constantly, Kelbell hopes he'll eventually begin to incorporate north into his mental gestalt. Subconsciously, he says, his brain will become tuned to the intermittent buzzing and non-buzzing, and he will know north—like a pigeon. Incidentally, that's behaviorism. It recalls the days of Pavlov and Skinner—the days of conditioning, and stimulus and response. Back then, at Harvard University, Skinner would put pigeons in his conditioning boxes and watch as they pushed levers to receive a delivery of seeds. But that's not the same as learning, assimilating, synthesizing, analyzing, recognizing, and understanding. I tell Kelbell I want to pilot a prototype device.

I meet Nadine Bonnett (1 out of 10) on the group page too. An Alberta resident, Bonnett was one of Iaria's study subjects at the University of Calgary. As a child in the Yukon, she remembers always struggling with spatial cognition. She has been lost all her life. Like many DTD sufferers in the group, she is reluctant to drive by herself, and she has countless stories of missed appointments, failed job interviews, and panicked hours spent in parking lots, searching for a misplaced car.

Reading posts from the group is a constant reminder that DTD is a complex disorder with unclear margins. Sometimes, it overlaps with other disorders. Group members might have any number of rare and mysterious syndromes that bring with them spatial disruptions—like Bálint's syndrome, a disorder that is accompanied by parietal cortex lesions; or Korsakoff's syndrome, which occurs with long-term thiamine deficiency; Schmahmann's syndrome, a condition that develops after cerebellar damage; or a genetic disorder like Turner syndrome.

Some members of the Facebook group report having something known as *visual reorientation illusions* (VRIs), the sense that your bearings are suddenly flipped by 90 or 180 degrees. Imagine that you walk out of a darkened movie theater and you're not where you thought you would be. A few might suffer from landmark agnosia: the very specific inability to recognize landmarks. Others might simply be suffering from *aphantasia*: the inability to mentally picture an object—literally, from the Greek, "without imagination." There is, however, a relatively straightforward test for it.*

Diagnosis of DTD is complex, and time consuming. It requires hours spent in an fMRI scanner performing spatial tasks, which is prohibitively costly. Most of us will never get the diagnosis we're looking for. Iaria tells me that he's currently not scanning subjects. He's waiting to hear back on grants he's submitted. I have missed the window. In the decades he's been researching DTD, Iaria has only taken around twenty people through the battery of behavioral tests, psychological evaluations, and brain scans required for a definitive diagnosis.

Even so, there is hope for the members of Directional Disorientation (aka Developmental Topographical Disorientation). In his lab, Iaria has been developing a computer-based training program designed to help people with DTD build a cognitive map. Published in 2020, results of preliminary studies with control subjects showed that the program improved the accuracy of the cognitive map they constructed. Iaria is hoping the program will help people like Patient One to form an inner map too.

•

Iaria has developed a battery of online tests that provides a window into how an individual processes spatial information.† Lacking the opportunity to climb into an fMRI scanner, the tests can be very informative. On a rainy Sunday afternoon, I sit at my computer and take them.

The first task is a facial recognition test. For several minutes, I stare at galleries of men's grainy black-and-white faces floating against a black background. For each question, I'm shown a face to study, first in three-quarter profile, looking off to one side; then head on; and then in three-quarter profile, staring off in the other direction. Then, I have

* Imagine a watering can. If you see a mental image of a watering can, you don't have aphantasia.

† The tests are here: info.gettinglost.ca.

to select the face from an identity parade of three similar faces. This test of facial recognition skills activates a different brain region, one not involved in spatial tasks—the fusiform face area. Unable to decode facial information, a prosopagnosic would fail the test miserably, but I get 72 out of 72 questions correct. A perfect score.

The next task is a test of my mental rotation abilities. This time, I'm shown an image of two three-dimensional objects made of blocks and oriented at different angles in space, and then asked the question *Are these objects the same or different?* In other words, can I study the first object and manipulate it in my mind, turning it in different planes to compare it with the second object? When I try, it's as if I can feel my brain stretching uncomfortably to perform these mental rotations. The answer sits just out of reach, like a musical note I can't quite sing. I score 64 out of 80. In part, the online script reads: *Your performance on this task indicates that you have an average ability to mentally rotate objects.* My score sounds like a good one, but since there are only two possible answers—*same* or *different*—even my dog, relying on chance alone, would be expected to score 40. Even so, Iaria tells me he doesn't necessarily expect someone with DTD to have serious issues with the mental rotation task.

My wife Emeline scores 74 out of 80. For her, the online script reads: *Your performance on this task indicates that you have an excellent ability to mentally rotate objects.*

Our scores seem relatively close, but it's the difference between comfort and discomfort—between effortless and effortful navigation. This is a task in which men typically outperform women. I ask my friends to take the test and one of them, Steve, scores 78, an almost perfect score. In April 2018, Nadine Bonnett underwent a full evaluation with Iaria, complete with fMRI scans. At home in Alberta, she has a three-dimensional model of her brain that Ford Burles, a scientist in Iaria's lab, printed for her. She sends me a photo of it: ivory-colored and grooved like coral. At first, she kept it on her shelf next to some houseplants, she says, until her three-year-old grandson became fascinated by it. Now it sits out of reach in a drawer instead.

Nadine scored 52 on the mental rotation task (*... you can mentally rotate objects with some difficulty*). From Nadine to me, to Emeline, to Steve, our scores—52, 64, 74, 78—tell the story of how well our brains allow us to process and manipulate space.

Next is the four mountains task. The test taker sees a computerized landscape—like what you might see far below you from the window of an airplane. It contains a flat plain with four round-shouldered lumps rising from it in a little hilly cluster. The mountains are not identical: one might be taller than the rest and lopsided, or one might rise to a sharp point like a pencil, or two of the mountains might be connected by a broad ridgeline. I'm then shown images of four similar landscapes. All of them contain four mountains, but only one of them shows the same topography as the first image, although seen from a different angle. It's a test of my ability to manipulate the mountains in my mind.

I score 9 out of 20. Emeline: 16. Again, it's the difference between struggle and ease. While I have to forcefully bend my mind to it, the mental manipulation is simple and seamless for her. With a score of 6—just slightly above chance—Nadine's results indicate that she has difficulty or an inability to remember or manipulate scenes in her mind.

"We're expecting some issues on the four mountains task," says Iaria, "because it's a task that is more difficult. It really requires your ability to recognize a complex scene from a different perspective."

The final test in Iaria's battery is known as the spatial configuration task. Five different shapes appear on the screen, floating serenely in the blackness of outer space: a ring, a cube, a spherical polyhedron, and so on. Iaria calls them, "five objects in infinite space." Their positions are fixed in relation to one another, and the test taker is asked to hold their configuration in mind. From the objects I can see on the screen, I must determine which object is immediately behind me. For instance, if I can see the floating cube, which is flanked on the left by the spherical polyhedron, then I must be in front of the pyramid.*

* This is a hypothetical example. I don't think I ever learned anything about how the shapes stood in relation to each other.

Again, the pattern of results holds. Steve scores 52 out of 60. Not far behind, Emeline scores 43. I am bewildered by the test. I try to force the clenched fist in my mind to open, fingers slowly unfurling like petals to reveal the answers. It doesn't happen. Even so, my score of 34 is better than Nadine's 21. And these numbers do the impossible. They allow us a glimpse into a complex system that contains millions of neurons firing in different structures across the brain—to appreciate the differences in how two people might view the world around them. They're informative in ways that a recording electrode implanted in a rat's brain, or an animal's movements in a maze, can never be.

Iaria checks my scores.

"Your performance on our tests, your description of your spatial difficulties, and the information that you report in our online assessment, are all consistent with what we'd expect from someone affected by DTD," he says. "All of them."

Chapter Eight

Your Brain, My Brain, His Brain, Her Brain

Ben Trumble is standing in the gloom of the Bolivian rainforest. It's an insect empire. Trumble, an anthropologist at Arizona State University (and an 8 out of 10), is tagging along with a group of Tsimane men—members of an isolated group of indigenous people who live in the Amazon basin. They live in settlements along the winding, tea-colored Maniqui River in lowland Bolivia. The black-haired, stocky Tsimane are foraging hunters. Riverine people. They still live a lot like we did 10,000 years ago. They use bows and arrows to shoot small monkeys from the trees and send their compact dogs into the undergrowth to retrieve the carcasses; if they have a supply of ammunition, they'll use a gun. They fish the river. In small, well-tended clearings in the jungle, they grow plantains, rice, and sweet cassava.

The forest canopy is alive with birdsong. Trumble watches as the hunters step nimbly between the trees. There are dangers here—among them, jaguars and other large predators, and plenty of venomous snakes. Trumble is in Bolivia with the Tsimane because he's interested in the physiological effects that hunting has on them.

"I was going out with hunters," he says, "and then I was taking saliva samples across the duration of the hunt, looking for key changes in testosterone and cortisol when they made a kill."

Periodically, during the hunt, he stops them. Standing in the shadows of the trees, he carefully collects their saliva, stores it away, and keeps moving. But, one day, Trumble noticed something else. "There was a lot of time following these hunters around and one of the things I asked them was, 'Do you know where we are right now?' " he says. He'd ask them, *Do you know where your village is*? Even miles from their home they could still point to it. It was as though they could feel it like a distant throb through the trees: *home . . . home . . . home.*

"I always thought it was just interesting how these guys found their way," says Trumble.

Navigation is vital to indigenous people like the Tsimane. There are no maps of the forest. It is a maze without walls. Regularly, the hunters disappear into the jungle, slipping between the tree trunks like shadows—and they will be gone for days at a time. Sometimes the Tsimane travel farther afield, walking miles through the forest to San Borja, a bustling market town with around 25,000 inhabitants. Back in the village, during his downtime, he began asking the Tsimane—all ages and both genders—to point to distant locations. At first, it was a game—a diversion from the boredom. But their way-finding ability was so impressive he realized he needed to study it. At five different settlements along the Maniqui River, he gave villagers a handheld GPS device and asked them to point it to three different places: the nearest upriver and downriver settlements, and the central plaza of San Borja.

The river is serpentine. It provides no navigational clues, even to the Tsimane who live along its banks. "If the river was straight it would probably take forty-five minutes in a boat to go from one of these communities to the market town," says Trumble. "Instead, it ends up taking six or eight hours, or maybe even a little bit more, just because it takes so many different twists and turns."

The forest is crisscrossed with footpaths that don't help the Tsimane get their bearings, either. The trees are dense and form a high, lofted canopy, like a green cathedral, that blocks sightlines to distant mountains and other landmarks. It blocks out the sun too. "It gets dark around three-thirty when the sun begins to set just a little bit," says Trumble. Instead, they must access their cognitive map of the forest and use path integration to chart their movement through the map. Standing under the canopy, they use their learned sense of where they are in relation to the other places in their spatial world. With the kind of intensive daily practice that only cultural traditions can provide, the Tsimane can perform this spatial feat even in low-light conditions, succeeding where Jan Souman's subjects, trying to maintain a straight line in the Bienwald forest, failed.

In his study, Trumble reported that Tsimane villagers can point to San Borja with an average error of around 25 degrees—and some perform even better than this. What does this actually mean? If San Borja was directly north but a villager pointed south instead, she would be exactly 180 degrees wrong. If, rather than pointing north, a villager pointed east, she'd be making an error of 90 degrees. An average error of 25 degrees is an uncanny level of accuracy. Some of the settlements were as far as fourteen miles overland from San Borja, and much farther by the winding river—a significant distance. Nevertheless, the Tsimane could point to it, unseen across tangled and trackless jungle, with incredible accuracy.

The women performed as well as the men. The young performed as well as the old—Trumble's oldest subject was eighty-two years old; his youngest was six. After reading the results of the study, I sit at my breakfast table with my sons—at the time, they're eleven, nine, and six—and I ask them to point toward three different and unseen locations: the building I work in; the elementary school they all attend; and the local park we visit all the time. If I'm asked to perform a task like this, I am always wrong—and usually by much more than 25 degrees. Sometimes by 180 degrees. The scientists call a woefully low success rate like that *below chance*. In other words, I might be slightly better at pointing in the right direction by spinning around three times and guessing. But all three sons get all three locations right. Sitting in the morning sun, crunching noisily on spoonsful of breakfast cereal, they can accurately jab a finger at my work building, even though it's three miles away on the other side of the city.

•

On a basic neuroanatomical level, I have the same brain as a Tsimane hunter. When I die, a pathologist will be able to remove my brain and compare it with the brain of one of the deadliest and most efficient Tsimane hunters—and discern almost no anatomical differences.

But even so, there *are* differences. And sometimes, those differences become significant. In fact, brains vary a lot between people. We tend to think of Einstein and Mozart as the outliers. But we're all outliers. No two brains are alike. And it's the subtle differences in our brain structures that make us who we are.

A few years ago, at Washington University in St. Louis, pediatric neurologist-in-training Nico Dosenbach formed something he calls the Midnight Scan Club. Motto: *carpe noctem*. Seize the night. Dosenbach wanted to solve a long-standing problem. In general, fMRI scans are low-resolution and full of noise. For most published fMRI brain studies, researchers scan the brains of lots of people—sometimes hundreds, and even thousands of individuals. The images are then averaged. They're combined. The end result is a *group* brain that scientists can use to make generalized statements about most brains. Sometimes it takes a lot of brains to say something meaningful about *the* brain. But, in reality, says Dosenbach, no one's brain really looks like the group brain.

For an analogy, Dosenbach says, consider people's faces. "Everybody's got a face that has many things in common," he says. "Most people have two eyes and eyebrows, and a nose, and all that. No two people's faces are the same."

Now imagine that we combine a thousand different human faces to form a single averaged version of a face: a group face. Averaging data from individual brains is just as misleading, says Dosenbach. "It's like taking all those faces and combining them, which honestly creates a cartoon face," he says. "Actually, that face doesn't look like a human face. If you gave an alien an image of that face and told it, *Look for things that look like that*, it would find cartoon characters, not real people."

As with faces, a group brain is just a cartoon brain. In some general ways, it's brain-like, but that doesn't make it informative in the same ways that a single brain can be informative. Dosenbach had already decided he wanted to explore the uniqueness represented by a single human brain, not the noise-filled combination of a thousand compiled brains. Instead of finding the ways brains are the same, he wanted to understand the ways they might differ from one other. The only way to do that is with high-density precision mapping of a single brain. As a junior researcher and low on funds, Dosenbach had heard that the cost of booking the fMRI scanner at Washington University was cut by around ninety percent after midnight, when it was not in use. At that moment, the Midnight Scan Club was formed. The club's Twitter page lists its six rules:

1. Get more data
2. Get more data
3. Scan at midnight
4. Don't move
5. Don't fall asleep
6. Share the data

As one of the first subjects, Dosenbach clambered into the scanner and lay restfully for hours, unmoving. He stared at a single white dot floating on a screen to induce a resting brain state. If he fell asleep or moved more than a fraction of a millimeter, the scan was ruined. For a total of $12,000, Dosenbach scanned the first ten brains—his own included. "In the neuroimaging research world," he says, "that's nothing. It's like pennies." The end result was ten high-resolution functional *connectomes*—detailed organizational maps, like wiring diagrams of the neural connections in each brain.

The data began to yield important findings straight away. For instance: not everyone's connectome is the same. Some people's brains seem to be wired completely differently. "In a sample of just ten, there were eight brains that looked the same-ish—or more similar," he says, "and then two that looked different, similar to each other on a fundamental organizational level."

Dosenbach's brain was one of the outliers. The connectomes of most of the other participants are organized in a loop—like a large, efficient circuit. Information travels quickest around an interconnected circuit like the brain by traveling in a loop. But in Dosenbach's brain, the loop is absent. As a result, his *global efficiency*—a specific measure of integrative information processing across remote brain regions—is slightly lower than in the other subjects. As yet, Dosenbach isn't sure what this means, how prevalent it is in the population, or the factors that determine whether an individual has a looped connectome or something different like his.

He's not the only scientist to show the uniqueness of every human brain, he's just shown it with superlative clarity. In a 2018 study, scientists scanned the brains of almost two hundred subjects three times across a two-year study period, converting different brain structures into a complex range of measurements. Neuroanatomically, each individual brain was so different that, by the end of the study, the researchers could identify each of the subjects based solely on their neuroanatomy—on the subtle architectural differences of certain brain regions. Our brains are as unique as our faces. And these individual brain differences are vitally important. As they accumulate, they make us who we are. These are the kinds of individual differences that Elissa Aminoff saw in J.N.'s brain and that Giuseppe Iaria sees in the brains of his DTD subjects when they perform spatial tasks in an fMRI scanner.

In 2009, the US National Institutes of Health launched the Human Connectome Project, a five-year-long study with the ambitious goal of mapping the human brain and its connections—like mapping a vast, complex, multilayered city. The study has provided new insights into the effects of brain structure on personality, and cognition, and behavior. In 2019, using data from the project, Dutch researchers showed that individual differences in brain structure helped to determine traits as distinct as working memory and language function. Other studies have even shown that brain structure impacts much more nebulous and subjective emotional states, like happiness and emotional well-being. For instance, a 2018 study of 1,000 people found that subjects with a relatively thin frontal cortex felt satisfied with life, but subjects with a thicker frontal cortex didn't. Another study has shown that connectivity between the right precuneus and the amygdala is positively associated with subjective happiness.

If the pathologist comparing my brain to a Tsimane hunter's brain looked closely enough, she might see that the hunter had a slightly larger hippocampal volume than I do, formed from decades in the rainforest—just like the enlarged hippocampus of a London taxi driver. Perhaps the hunter's white matter tracts, which connect distant brain regions to one another, are more voluminous, or sturdier, or take a slightly more direct path through his brain than they do through mine. Maybe certain cortical regions are grooved and ridged a little differently than mine, and are more efficient at processing the complex spatial information he uses to make sense of his world. The differences will be subtle. But he uses his brain differently, and that's because it *is* different.

All human brains are similar, but only in the same way that the cities of Mecca in Saudi Arabia, Havana in Cuba, and Brussels in Belgium are similar. They're all large cities with around two million inhabitants. But Brussels is no Mecca, and vice versa. This is how science sometimes fails us. It ignores the outliers, because they're inconvenient. Whenever I've asked scientists why some people, like my wife, are good at navigating while others, like me, are terrible at it, they have all said they don't know. But, at least in a very simplified way, we *do* already know. For instance, we know that we're all born with endlessly complicated brains. But they're complicated in different ways. This is a notion that scientists hadn't really appreciated or even discerned until a few decades ago. And it has taken recent work like Dosenbach's Midnight Scan Club and other studies like the Human Connectome Project to really show it clearly. We know that life then alters our brains even more. Everything changes it.

Culture changes it. Language changes it. Origami changes it. Even the presence or absence of light changes it.

Navigation is one of the most cognitively complex tasks our brain performs. It depends on the exchange of information between lots of distinct brain structures, all of them slightly different in different people. Information barrels from one brain region to another, flickering across the cortical surface like an electrical storm. And all of the involved regions—even the serpentine folding patterns on the brain's surface—are slightly different in each of us. Is it a surprise that some people might be better at navigation than others? It's the same reason some people are better at assessing risk, or empathizing with others, or speaking multiple languages, or playing a musical instrument, or solving complex mathematical equations, or even being optimistic.

It's easy to imagine that a system like this produces people like me, and people like my wife. It's a system that sends Amanda Eller farther into the tangled green depths of the Maui rainforest instead of walking back to her parked car. The more parts there are to a system, the likelier it is to fail, especially when every single component is unique, partly shaped by culture and experience, and partly inherited.

After all, hammers don't fail. Complex machines fail.

•

What causes these differences in brain structure and function? Kaili Rimfeld (7 out of 10), a statistical geneticist at King's College London, is trying to find out. Rimfeld's previous research has focused mostly on the genetic factors that contribute to educational achievement. More recently, she began asking a slightly different question: to what extent are our spatial skills decided by the genes passed on to us by our parents? What do we inherit? What is determined instead by our environment?

In a series of studies, Rimfeld subjected 1,300 twin pairs to a battery of spatial tests assessing abilities like mental rotation, map reading, route memory, and navigation. "The twin method offers this really nice experiment to us," says Rimfeld, "because we have monozygotic twins who are identical to each other, and we have dizygotic twins, who share fifty percent of the segregating genes—so, the genes that differ between individuals." With this approach, geneticists like Rimfeld can begin to tease apart the separate effects of genes and environment that go to make up a particular trait.

"If the monozygotic correlations are higher than the dizygotic correlations, then we can already infer that there is genetic influence on these traits," she says. At the start of the study, Rimfeld had assumed that spatial abilities could be separated into their component parts. For instance, someone might be proficient at one aspect of it, like map reading, but struggle with another, such as mental rotation, or spatial reasoning. Not so, she says. All the separate components of navigation seem to cluster together into a single factor, which she calls, predictably enough, *spatial ability*.

"There is no separable navigation factor, or mental rotation factor, or visualization factor," she says. In other words, if you struggle with one component of spatial ability, you probably struggle with all of them. If your brain can mentally rotate objects, you're probably proficient at map reading and memorizing a route. "On average in a population, if you're good at one thing you usually are good at the other thing as well," she says. From her 2020 twin study, Rimfeld found that around 84 percent of the variance in our spatial abilities is decided genetically. That seemed like a lot to me. But not to Rimfeld.

"It's not surprising at all, actually," she says. "We've known for a long time already that cognitive abilities are highly heritable." In other words, says Rimfeld, individual differences in spatial abilities, just like cognitive abilities, are largely explained by genetic factors. But just how remains complicated. For instance, as many as fifteen different genes determine eye color—a relatively straightforward genetic trait. The more complex the characteristic is—like spatial ability or educational attainment—the greater the number of genes that are involved.

"The latest GWAS—that's genome-wide association study—used 1.3 million people and then identified over 1,200 genetic variants that are associated with educational attainment," says Rimfeld. She expects spatial ability involves thousands of gene variants too, each contributing an almost imperceptible effect to the whole. It's a complex trait, built gene by gene.

Let's reconsider the pedigree of French orienteering champion Thierry Gueorgiou, who dominated the sport worldwide for more than a decade in the early 2000s. In a YouTube video from 2014, you can watch him compete in Portugal. Map in hand, he runs toward checkpoint number 5—5 of 29 on the course—threading a path between pine trees and scrub. His gait is efficient: slightly hunched over, elbows out, long-legged strides in the sun. Gueorgiou consults the map often, never breaking his stride across the unfamiliar terrain. As he snakes between the trees, an inset map shows the course and Gueorgiou's current position on it, updated in real time by GPS. He is a straight blue line streaking directly toward the checkpoint. The cameraman can't keep up—his shadow zips across the uneven ground as Gueorgiou disappears in the distance. He won the race, finishing almost two minutes ahead of Swede Albin Ridefelt. A few months later in Italy, Gueorgiou won the men's long-distance event at the world orienteering championships.

When I ask him why he's such a deft navigator, Gueorgiou tells me his father coached the French national orienteering team, and his mother was the president of the local orienteering club; his brother was also a member of the French national team; his partner, Annika Billstam, is a three-time orienteering world champion who competes for Sweden. Together, they have a son and a daughter, currently aged four and two. Maybe the combinations of their parents' colliding genes will create spatial powerhouses.

Recently, I asked my mother to rate her navigation abilities on a scale from 1 to 10, with 1 being very bad. We know from Mary Hegarty's Santa Barbara Sense of Direction scale that an individual's self-assessment of spatial abilities is accurate. My mother gave herself a rating of 1: very bad. With his family genetic studies of DTD, Giuseppe Iaria determined that everyone with DTD has at least one parent with it. On the other hand, my father has an almost supernatural ability to navigate. When I was young, it seemed as if all the streets of England were known to him: the newly built '70s-era ring roads and bypasses; the ancient repaved

Roman roads; the motorways, the one-car-wide country lanes that appear on maps as sinuous lines as thin as capillaries, if they appear at all. But I never saw him *look* at a map. He just knew it.

On the genetic battlefield, thousands of genes collided with one another, like two rivers meeting. My mother's genes won. On balance, I'm satisfied with the results of the collision. My sister rates herself a 5. In her case, the battle ended in a draw. My nonidentical twin brother does not get lost on his own street. The thousands of gene variants that determined his spatial abilities just collided in his favor.

In the future, Rimfeld plans to start identifying some of the thousands of candidate genes involved in spatial cognition—a difficult task. As with the studies to decode the genes involved in educational attainment, it will take millions of subjects, all taking the same tests under the same exact conditions, to detect the signal in the noise.

"Right now we just know that genetic variants are involved, but which genetic variants and in what way, we just simply don't know," she says. "To get to the needed sample sizes of a million or two is just really, really hard."

•

Even so, a brain isn't a static thing. The brain I was born with forty-eight years ago has been steadily altered during the decades I've been (occasionally) using it. In fact, the brain changes constantly, altered by what we ask it to do—a concept known as neuroplasticity. For someone with poor spatial cognition, says Rimfeld, a carefully designed intervention has the potential to improve navigation abilities. In this way, future interventions might allow us to increase the variance in our spatial abilities determined by the environment—the 16 percent in Rimfeld's study—and try to decrease the much larger genetic contribution.

Research has identified lots of activities that can alter the brain, remodeling it from within. In Vienna, in 1781, when he was twenty-five years old, Wolfgang Amadeus Mozart wrote the Sonata for Two Pianos in D Major. Across three movements, two pianists play complex melodies that spiral and climb, sometimes coming together briefly to form playful harmonies before flying apart again at an insistent gallop. Resolving. Separating. Rejoining, woven together. Flying apart again. It's a pretty piece.

In 1993, researchers claimed it was responsible for something known as the Mozart effect. After listening to the Mozart piece for ten minutes, scientists reported, college students performed spatial tasks better than students who had listened either to a relaxation tape designed to lower their blood pressure, or nothing at all. For a few weeks last spring, when I walked my dog in the woods, I did it to Mozart's sonata. Every day, we walked through a carpet of bright green sapling maples and jacks-in-the-pulpit to the sound of two pianos, their melodies intertwining, coming together, and then flying apart again, like the birds half-seen in the branches on either side of us. And every day, at some point, as the two entwined melodies separated, I walked until I was in the forgotten, overgrown, too-quiet part of the woods, where someone might decide to hide a body.

The Mozart effect is controversial now. Since the original paper was published in 1993, other researchers have struggled again and again to replicate its findings. In every other study, when people listen to the sonata, nothing happens. A later study found no difference in spatial skills whether subjects listened to Mozart, or Philip Glass, or silence. Another compared Mozart, Bach, and silence: still no effect. The scientists who described the Mozart effect went on to claim that rats performed better in a maze as adults if they heard Mozart as fetuses. They take fewer dead ends. They're faster to the reward than rats that had only heard white noise, or the music of Philip Glass, *in utero*. Appalachian State University psychologist Kenneth Steele (8 out of 10) wasn't convinced. He has spent decades carefully—and convincingly—debunking the Mozart effect. Others have too. A 2010 study in the journal *Intelligence* was titled "Mozart Effect–Shmozart Effect: A Meta-Analysis."

Mozart aside, the science is clear: we can change our brains. Since 1993, several other studies have found a positive cognitive effect of listening to music, as long as the listener finds it enjoyable. Implausibly, if you enjoy Barry Manilow, you'll benefit from listening to "Copacabana." Or, as Steele says to me, "The bottom line is that listening to Mozart while trying to navigate the woods will offer no differential improvement in your navigational skills compared to listening to Philip Glass or Insane Clown Posse."

Other activities affect the brain too. For instance, expert musicians have a larger corpus callosum, the thick, white-matter nerve tract that connects the two hemispheres of the brain. Jugglers show increases in grey matter in the brain regions that process visual motion. A 2019 study showed that being multilingual actually alters the architecture of the brain.

In 2016, researchers at Carnegie Mellon University trained subjects to learn a virtual spatial route. After less than an hour of training, they saw in their subject's brains an increased connectivity between the hippocampus and the right intraparietal sulcus—a brain region involved in navigation—and the posterior temporal areas, which are activated by spatial memory tasks. In real time, they observed the brain making new neural connections, remodeling and altering itself.

Perhaps there are other ways to change a brain too. For a 2018 study, researchers gave Siberian chipmunks two different substances: docosahexaenoic acid (DHA), an omega-3 fatty acid essential for brain development, and uridine-5-monophosphate (UMP), one of the building blocks of RNA. At certain times, spatial cognition is very important to a Siberian chipmunk. In winter, like squirrels and other hoarding species, Siberian chipmunks rely on the food stores they hid underground earlier in the year, when food was plentiful. In leaner times, they will starve if they can't remember where they buried their hoards. After just six weeks of orally administered DHA and UMP, chipmunks store more seeds than control animals. They perform better in a Y-maze too. They become spatial superstars. Suddenly, they're the Mozarts of food hoarding.

Within a few weeks of receiving DHA and UMP, Siberian chipmunks have a measurably larger hippocampal volume. Their brains are altered by it. In other words, perhaps it's possible to get the brain of a London cabbie without spending years bent over a street map.

•

I live by the general principle that nothing is free. It's a principle that holds for neuroplasticity too. A London cabbie can stuff his posterior hippocampus with so much spatial information that it grows in size, occupying more space in the brain. But those changes come with a cost.

Since a taxi driver's hippocampus is an enclosed structure nestled deep in the center of the brain, it can't just expand infinitely. There are limits. So, while the posterior part of a cabbie's hippocampus—the part involved in spatial memory and navigation—gets larger, the anterior part of it, located toward the front of the brain, becomes smaller.

And that matters. Taxi drivers score much better at tests like the "London landmark recognition memory test," which, as the name suggests, measures their memory for landmarks. They're also better than control subjects at the "London landmark proximity judgements test," in which participants are shown images of a target landmark and two other landmarks and asked to decide which of the latter is closest to the target landmark. They can interrogate their cognitive map for the answer. But if you show them a complex image to memorize and then redraw, they're worse. They also score worse than control subjects on the "verbal paired associates test"—a test of episodic memory. They are master navigators, but at the expense of other abilities—in this case, at the expense of associative memory.

Their brains are plastic, but not infinitely so.

•

What we inherit from our parents doesn't explain all of our spatial abilities. Neither do our individual efforts to change our brains—by juggling, or studying road maps, or by listening to Insane Clown Posse. The culture that we're part of has an important effect too. Different cultures approach navigation differently. If you drop me in the Bolivian rainforest with the Tsimane and ask me to navigate, I will fail terribly. But so will my wife, the master navigator. The Tsimane have been prepared for it in important ways by their culture. Think of the brain as a net, and culture as the water that passes through it.

One of the most powerful aspects of culture is language—now, just as it was for early humans as they competed with Neanderthals for available resources. The way a particular culture understands space is shaped by the language we use to describe it. When we talk about an environment, we use frames of reference as a way of defining space and the locations of objects within it. We can use several different frames of reference, and they differ in important ways.

For instance, there is the *relative* frame of reference, which is egocentric and based on our own position in space. If I'm using the relative frame of reference, I might say, *My coffee mug is to the right of the kitchen sink*. In other words, I describe the location of the mug based on my own position, and relative to another object. If I instead use the *intrinsic* frame of reference, I would describe the same situation by saying *My coffee mug is to* my *right*. I don't rely on another object to describe my mug's location. Finally, there's also the *absolute* frame of reference, which relies on fixed bearings like cardinal directions—north, east, south, and west. Generally, we use the absolute frame of reference to describe locations in large-scale environments. I would never say, *My coffee mug is southeast of my kitchen sink*.

But, since some cultures think about space differently, they do describe small-scale spaces like that. An Aboriginal Australian from northern Queensland who speaks the Guugu Yimithirr language has no descriptive terms like *left* and *right*, or *in front of*. Instead, she will say, *You are holding your coffee mug in your south hand*. Guugu Yimithirr speakers are unusual, but not unique. Tenejapan Mayans, who live in the highland region of Chiapas, Mexico, use terms like *uphill*, *downhill*, and *cross-hill* to describe locations in their hilly environment. Instead of saying, *Please pass the coffee mug that is to the left of the kitchen sink*, they will ask, *Please pass the cup that is uphill*. The Tenejapan Mayans, who speak Tzeltal, are incredibly accurate pointers too—like Guugu Yimithirr speakers. Every time a Tzeltal speaker even thinks briefly about location—or anything else, for that matter—that thought has an uphill, or downhill, or cross-hill component to it. Everything is spatial. By some estimates, around 10 percent of all words spoken in a Guugu Yimithirr conversation are either *north*, *east*, *south*, or *west*.

The bird flew into the north window.
The boy fell off his bike on the southwest road.

•

Research shows that how we talk about space determines how our brains understand and process space too. From the moment we're born, our brains are shaped by the culture we're part of. "It doesn't seem surprising that culture should shape the brain," says University of Utah anthropologist Elizabeth Cashdan. "How could it not?"

Cashdan (who says she's "a 4 on a good day") has studied spatial cognition in several different populations—in the Tsimane in Bolivia, as well as in Namibians, Yucatec Mayans, and the Hadza, a group of modern hunter-gatherers in northern Tanzania. Different environments, she says, lead to different spatial requirements, and that results in different kinds of navigation. A Tsimane hunter who is used to traveling by forest paths relies on different techniques to navigate than someone like Cashdan, who lives in Salt Lake City, a city built on an orderly and geometric grid, bordered on one side by a mountain range, in pan-flat Salt Lake Valley, Utah. Consequently, Cashdan says, she's "a lousy navigator."

Our culture did this to us. As Hugo Spiers's *Sea Hero Quest* data have shown, in some countries men and women live very different cultural lives. Their brains are shaped by different pressures and experiences, with women prohibited from driving and prevented from moving freely in their environment—prevented, too, from constructing dependable cognitive maps. Is it any surprise that different cultures lead to different spatial abilities?

"I assume that you and I would be better navigators if we grew up in an environment with a lot of navigational challenges," says Cashdan, "and perhaps if we had a language that forced us to refer to locations using cardinal directions."

•

It doesn't take years of studying a London road map to change the brain. Recent research suggests that the brain is fundamentally altered by even the smallest and least consequential events and experiences. Researchers at Harvard Medical School have been trying to determine precisely how the environment alters brain function. For a 2018 study, they raised mice in total darkness from birth; then, after exposing the mice to light, they measured the effect of the light on the expression of different genes in thousands of individual neurons in the brain's visual cortex, comparing it to neurons in mice that were never exposed to light.

The results were surprising: the expression of thousands of different genes was altered by a moment of light—a brief flash. Light is just one environmental factor. It's likely that genes are turned on and off in the brain by all kinds of events and experiences. In turn, the activity of neurons and entire complex neural networks is altered by environmental factors. I ask Harvard University's Aurel Nagy, one of the study authors, what this really means. For instance, what other factors in addition to light might alter brains?

Most likely, he says, everything we encounter has the potential to subtly change our brains. *Everything* changes the brain. Nagy speculates that we all start out with a basic pattern of connections—consider it a blank slate, he says. This pattern will differ from one individual to another. But then experiences begin to alter that blank slate, first via gene expression changes, which then lay the molecular groundwork for longer term structural and connectome changes in the brain. When we're young, he says, and new connections are forming, we're more easily molded by our surroundings.

Consider the implications of Nagy's research: my wife and I both began with a blank slate, a basic connectome. From that point on, every experience and event subtly altered it, cultivated, shaped, and remodeled it. Genes were turned on or off. Millions of neural connections were made, synapses bridged, impulses fired, the cortex enfolded in certain ways. And the same is true for all humans: for Amanda Eller wandering the Makawao Forest Reserve, for London taxi drivers, for Tsimane children; for Tzeltal speakers, born in a simple house on a cross-hill street in the Chiapas highlands; for the unliberated women of Pakistan and India and Saudi Arabia.

•

The Tsimane aren't born with superior navigation skills. They earn them. It's unclear whether they are born with a genetic advantage, or if their connectomes are different—no one is transporting an MRI scanner into the Bolivian lowlands. But they still have to develop their abilities. And they do it the hard way: by getting lost in the forest. In fact, being lost is an unavoidable and necessary part of their life experience. The average hunting trip, says anthropologist Ben Trumble, involves a nine-hour walk and covers a distance of more than ten miles. Every hunter he spoke with had a story about the time he was lost in the wild, far from home, as the light began to wane. Inevitably, they lose their bearings beneath the canopy. It's dark. They head in the wrong direction, into a green tangle that hides jaguars and caiman and other predators. Sometimes, they're lost for days at a time. Usually, they find a familiar trail at some point. They limp home.

"Some of them would tell these horrendous stories where they were lost for five days and ended up completely on the other side of the Tsimane territory," says Trumble. "There was a little boy recently who got lost, and he was gone for, I think, three days before he was found."

And sometimes, despite their superior navigation abilities, they die out there. The rainforest swallows them whole. The distant throb of their home becomes too faint and they can no longer hear it filtering through the trees. For the Tsimane, getting lost and dying in an unfamiliar part of the jungle is a real possibility. In some indigenous groups, Trumble says, predation can be responsible for almost 10 percent of male mortality in certain age groups. In other words, hikers on Maui can get lost in a forest reserve and suffer very bad outcomes, but so can people who spend their entire lives in the wilderness. Importantly, there are always outliers: some of Trumble's Tsimane subjects just weren't accurate when pointing at distant unseen locations across the rainforest. In fact, they were very bad at it. The bell curve exists in the rainforest too, and inevitably, some people will reside in the tails of the curve, either exceptional at navigating, or terrible—10s or 1s.

"I remember checking to make sure it wasn't just younger people who were really terrible at it because they hadn't traveled as much," he says. "Or that it was older people who were perhaps in early stages of cognitive decline. There isn't any evidence of any sort of bias in that direction."

There was no pattern to the failure. "There happened to be some people who weren't so great at it," he says. Despite their inherited gifts, and the benefits their culture has given them, and the lessons learned (and sometimes not learned) from living in the rainforest, they get lost.

•

The Tsimane are not unique. For thousands of years, the Inuit hunters of Igloolik have navigated an even more challenging physical environment. A small hamlet on Igloolik Island in Nunavut, northern Canada, the town of Igloolik is perched on the northernmost edge of the world. The region is surrounded by ice floes, frequently locked in by fog, and pummeled by whiteouts that can last for days. This is the world that Ferdinand von Wrangell described two hundred years ago, and Charles Darwin used to formulate his ideas about path integration in 1873. For most of the year, Igloolik resembles a blank and unlined sheet of paper. Nevertheless, the Inuit manage to move across it—across the blank page—with purpose and intent.

As anthropologist Claudio Aporta put it in a 2005 study of Inuit navigation: "Difficult by any measure, traveling in this land- and seascape as the Inuit people have done for millennia requires a sophisticated knowledge of subtle qualities: coastline shape, stone cairns, snowdrifts, wind direction, currents, animal movements, dreams, and other clues." There is something uncontainable in this statement. It belongs to the magical world. Dreams and other clues. If the Inuit rely on dreams and other clues to navigate their icy world, what chance would I have? I get lost on my own street. How could I use stone cairns and snowdrifts to find my way in Igloolik? This, again, is the power of culture. For his research, Aporta (a 6 out of 10) accompanied Inuit hunters on a walrus hunt in impenetrable fog. Like an illusion, the boat gently seesawed in a silent nothingness, a boat floating through an endless cloud. Somewhere deep beneath them, walruses corkscrewed slowly through the water, invisible.

"There were times when we could see absolutely nothing, just because of the fog," says Aporta. Regardless, the Inuit always knew where they were. In fact, says Aporta, a skilled hunter was never truly lost. At times, they might get what he calls temporarily misplaced. They navigate by tidal movements, and rhythms of the water, and subtle shifts in the current. At other times, they might monitor the shape of snowdrifts, and the spatial patterns of ridges, or boulders. As he hunted with the Inuit, and traveled with them through their landscape, Aporta became more aware of the cultural aspects of navigation.

"There was a social and historical dimension to the landscape that was completely out of my perception," he says. "I wasn't traveling in a no-man's-land. I was traveling on a social landscape. The one thing that surprised me the most was their ability to remember so many features in an area that was very large. They could actually describe, you know, a rock, and the color of the rock, and how the rock would look depending on where you were. That was fascinating."

In a place that seems featureless, the Inuit see information everywhere—in a rock, or a wind-bent tree, or the particular shape and movement of a floating piece of sea ice.

"In terms of getting hopelessly lost in the Arctic," he says, "that doesn't really happen to experienced people. They have so many ways of establishing their own locations and reading the landscape, and features, and the sea ice and so on that it really could not happen to a very skilled Inuit."

But, says Aporta, that's changing. And it's changing fast.

Chapter Nine

The Future

Laugavegur.

That's what Noel Santillan had intended to type into the rental car's GPS system. It was February 2016, and he was on his way to Hótel Frón, where he'd booked a room for a week. The hotel is located on Laugavegur, a main business artery in downtown Reykjavik, the capital city of Iceland. An hour earlier, Santillan, a twenty-eight-year-old retail manager from New Jersey who rates his spatial skills at 8 out of 10, had landed at Iceland's Keflavik International Airport and loaded his bags into the back of his rental car, a Nissan. The hotel was a forty-minute journey, northward from the airport into the heart of the downtown Reykjavik—or "Bay of Smokes," in Icelandic.

But, instead, Santillan had entered *Laugarvegur.* If you look closely, you'll see there's an extra and unneeded *r* in the middle of it. That *r* was going to become a problem. In the Icelandic language, *Laugavegur* is an old word meaning "wash road," as in: the road where people take their laundry to wash it. But, with its extra *r*, *Laugarvegur* doesn't mean anything at all. If you Google *Laugarvegur*, you'll just find the story of what happened to Santillan next.

For some reason, the rental car's GPS responded strangely to the typographical error. It rerouted him instead to a town called Siglufjörður. In a 2017 article about Santillan, *Iceland Magazine* did not report this part of the story as an unexpected development in the same way I think of it as an unexpected development. Perhaps there's something uniquely Icelandic and understandable about it.

Laugavegur.

Laugarvegur.

Siglufjörður.

Santillan hadn't noticed. The scenery in Iceland is otherworldly—commanding enough that he can be forgiven, jet-lagged and wired from the flight, for not noticing the error. The rental car zipped northeast across an icy planet, black rocks on either side of the road, vents issuing little columns of volcanic steam as if the entire country sat on top of a furnace. On his left, the sea was slate grey and looked hard, like cement.

I've Googled the drive from Keflavik Airport to Hótel Frón: a short drive on a highway that mostly hugs the rocky, wind-beaten coastline, past a museum called *Viking World*, and several geothermal beaches, before it hits the outskirts of the city. (The hotel sits about four blocks from the Icelandic Phallological Museum, which boasts a collection of more than 200 penises and penis parts representing almost all Icelandic sea and land mammals. Google says it's "family-friendly.") I spend a few minutes looking at photos of the rooms at Hótel Frón too: clean and practical—and irrelevant now. Noel wasn't going there anymore. He wasn't going to the penis museum either. He was going to Siglufjörður.

The problem: Siglufjörður was nowhere near Santillan's hotel. Instead, it's a little fishing town nestled in the crook of an isolated bay, at the foot of steep green mountains, on Iceland's northern coast. With around 1,200 inhabitants, Siglufjörður is a world away from the neat and tidy rooms at Hótel Frón. It has its own museum, called the Herring Era Museum, and not much else. Piloting the Nissan carefully north along icy roads, the drive took Santillan more than five hours. But he never stopped to ask himself why it was taking so long to get there.

•

Santillan's story is entertaining, but not unique. After his week in Iceland, he became a minor celebrity. It's an entertaining story because Santillan is still alive. Other people have been less fortunate. The fate they suffered has a name: *Death by GPS*. Death by GPS is common enough that it has its own entry on Wikipedia.

For instance, in March 2015, Iftikhar Hussain's GPS sent him driving his Nissan Sentra across a bridge in Indiana that was no longer there. An elevated highway span had been dismantled in 2009—ending abruptly almost forty feet above the ground, like a cartoon bridge. Hussain was severely injured when the car plunged to the ground and burst into flames. His wife died.

Another unfortunate example: in early 2011, Albert and Rita Chretien were driving their minivan south from British Columbia to Las Vegas when their GPS sent them along a remote and barely-there road in Nevada. On the GPS it was a bold line that demanded to be driven on. In reality, it was an unmaintained road into the mountains. Eventually, they became stranded. Albert left the minivan on foot and died wading through deep snow. Rita was found with the minivan, almost dead, forty-eight days later, after surviving on snow and trail mix. They had simply followed the instructions.

Even when it doesn't involve death, a glitch in the system can cause mayhem. In June 2019, Google Maps diverted drivers around a blocked road near Denver International Airport. The detour sent almost 100 drivers into a field, where the cars fishtailed and eventually became stranded in mud.

In 2009, a vacationing Swedish couple also had trouble with a misplaced *r*. Intending to type Capri into their GPS, they typed Carpi instead.

Capri.

Carpi.

Capri is an upscale island destination in the Bay of Naples, floating in a glittering turquoise sea midway along the length of Italy's western coast. Sleek luxury yachts sit in the harbor like spaceships. Celebrities stroll photogenically along the charming cobbled streets. But you won't see that in Carpi, a mostly industrial town 400 miles away in the hilly northern province of Modena. While it looks charming in photos, it's not an island in the Bay of Naples. You won't see a sun-kissed DiCaprio there. In fact, some of the most charming images of Carpi that result from a Google search are actually just images of a misspelled Capri, proving how easy it is to misplace an *r*. (For instance, landlocked Carpi doesn't have a busy marina, with boats bobbing up and down on the turquoise water. But sometimes, because of misspellings, Google thinks it does.)

•

Computer scientist Allen Lin (9 out of 10) studied the patterns in deaths by GPS while at Northwestern University. He now works for Google. In a grim form of accounting, Lin built a database of more than 150 news articles that detail catastrophic GPS-related incidents. He collects the deaths. The plunges. The disasters.

There's a litany of them. Incident: the Japanese tourist who unwittingly drove into the ocean, guided by a GPS unit. Incident: the Belgian woman who left her home near Brussels for a train station forty miles away but drove 900 miles across Europe, almost all the way to Zagreb in Croatia, under the irresistible spell of a broken GPS device. Incident: the group of skiers who intended to drive to La Plagne, a ski resort located high on the slopes of the French Alps, but instead arrived at Plagne in southern France—almost 500 miles from La Plagne.

La Plagne

Plagne

"There are two main types of things that we found that were the root cause of these Death by GPS accidents," says Lin. "One big category is the incidents that are related to the algorithms and the systems themselves."

More than half of the incidents Lin collected were caused by missing or incorrect road attribute data in GPS databases, he says. While personal navigation systems often include exhaustive geometries of the road network, they'll lack important data about its physical characteristics. The clearance height of a bridge is wrong. A ferry line appears mistakenly as a road. Whether or not a road crosses the border of a country is listed incorrectly. Incident: a woman follows a route faithfully, driving into a swamp in Ontario, and is later found standing on top of her submerged car. Incident: a GPS device instructs a driver to enter a freeway via the exit ramp and she proceeds to drive on the wrong side of the highway for thirty miles.

Occasionally, drivers are instructed to turn onto roads that are unpaved, or no longer maintained, or that had been planned and appeared on maps but were never actually finished. "We saw a lot of instances in which people drive a two-wheel-drive car, but the GPS routes them to the shortest path, which happens to be a dirt road," says Lin. "Add some weather constraints and they're trapped on the road for hours or days." Incident: a medical student follows her GPS onto a remote logging road in New Brunswick, Canada, and is stranded in snow for three days.

"Another category of cause is related to the human-computer interaction part," says Lin. For instance, the cartography of a GPS device usually deemphasizes objects that aren't road networks. This is an attempt to reduce the driver's information load—to focus on the signal instead of the noise. But sometimes, says Lin, a feature isn't deemphasized enough, allowing complex geographies to introduce unexpected confusion. If a railway track sits close enough to a road and is visible on a map, people will mistake it for a road and drive onto the track.

Roadway

Railway

More than half the incidents Lin collected in his database involved collisions. Most of them were single-vehicle collisions. A fifth of the incidents left drivers stranded in rural environments, stuck in the wilderness as the light failed, Googling *wolves* on a half-charged cell phone. A quarter of all incidents involved a death, or several deaths. For the last several years, when newly built cars roll off most manufacturer's production lots, a standard model comes equipped with GPS. When we totally give ourselves to it—when we let our critical faculties go dark and drive across Europe to Zagreb instead of to the nearby train station—we are in trouble.

•

Convenience comes at a cost. As previous research has shown, a London taxi driver's enlarged posterior hippocampus is a trade-off that might give him superlative navigation skills at the expense of his working memory. When we opt to use a GPS device, we're no longer using our own powerful internal navigation system. Eventually, with neglect and disuse, the hippocampus becomes nonresponsive.

Anthropologist Claudio Aporta has seen this firsthand among the skilled Inuit hunters he has studied in Igloolik, in northern Canada. In a 2005 paper in *Current Anthropology*, he tells a story: he'd met a seventeen-year-old friend outside a grocery store in Igloolik. The friend was panicking. His father had asked him to retrieve a broken-down snowmobile that he'd abandoned several days earlier a few kilometers from town. The son had been searching for it—now undoubtedly covered by a thick blanket of snow, as everything eventually is in Igloolik.

But he couldn't find it. A generation earlier, a father would have been able to guide his son to the precise geographic location of a snow-buried snowmobile, sending him through the snow. In a way, it was ironic that the son was searching for a snowmobile encased in snow. The gradual erosion of Inuit way-finding abilities owes a lot to the modern and widespread use of GPS devices—but it began with the arrival of snowmobiles.

Go back far enough (not even very long ago) and the Inuit traveled by dogsled. It's elemental: a wooden sled pulled by a team of dogs. It was slow—or slow relative to a snowmobile. Aboard a sled, the Inuit could observe the landscape and its features and subtleties as it slid past them.

"That ability is diminished when traveling on snowmobiles," says Aporta. "When you travel by dogsled the pace is slower. You have more time to look around and, physically, you *can* actually look around."

Older Inuit had navigated their sleds across the snow and pack ice via well-known routes. Traveling by sled, the breath of the dogs trailing behind in a long steamy cloud, they had both the time and opportunity to discern the particular realities of a certain rock or a broad-shouldered hillock, or the shape and angle of a wind-sculpted snowdrift. Try doing that from a snowmobile as you hurtle across the snow—across the blank and unlined page. It's impossible.

Inuit elders would tell members of the younger generations to make sure they practiced not only looking in the direction they were traveling, but turning back periodically and looking at where they'd come from too, and surveying the landscape as it unspooled on either side of them. This is how a cognitive map is constructed, after all—one data point at a time, the margins slowly extending, the interior gradually becoming populated with information. Not by zipping across the snow at fifty miles an hour hunched on the back of a snowmobile, eyes locked on the horizon.

The snowmobiles came first. They were the first trickle of what later became a technological torrent. "GPS came into people's lives after they had embraced snowmobiling," says Aporta. "Snowmobiling, you know, you go faster, you have less time to look around, so the GPS becomes part of that assemblage of devices."

Around the same time, the Inuit came in off the ice. They began to move into modern settlements, which gave them access to batteries, computers, and lots of other kinds of technology. "Those technologies become entangled in people's lives," says Aporta.

The fall had begun.

When the Inuit father instructed his son to find the disabled and buried snowmobile on the outskirts of town, says Aporta, he gave the son oral directions using the old Inuit place names—the names that get passed down and don't appear on maps. He also used references to things like wind direction. He expected his son to be able to enter the strange, half-real dream space of intuition and snowdrifts and rocks with particular and memorable shapes. But the names and the wind direction meant nothing to the son—a modern son, born in a settlement.

Aporta, who had been with the father days earlier when the snowmobile broke down, had recorded the location of the snowmobile with a GPS receiver. The next day, with a freshly printed map in hand, the son retrieved the snowmobile.

•

In a 2020 study, Harvard College of Medicine researcher Louisa Dahmani decided to measure the effects of GPS on spatial memory. When she moved to Massachusetts from Canada and first encountered Cambridge and its winding, unpredictable streets, Dahmani, a previously confident navigator, realized she hadn't been tested by her environment before, and downgraded her self-assessment score to 6 out of 10.

For her study, she recruited fifty drivers, and tested their navigation abilities with two different virtual navigation tasks. They made their way through computerized radial arm mazes. In the distance, distinctive peaks and little pointed hillocks marked the horizon acting as virtual landmarks. "We also approximated the total number of hours of GPS use over their lifetime," Dahmani told me. She wanted to know whether people who relied on GPS navigated differently than those who didn't.

"We found that the more people used GPS," she says, "the more they relied on navigation strategies that mimic the GPS. They mostly ignore the layout of their environment, like where various landmarks are, and instead they learn a series of directions to get to a specific destination."

During the virtual task, an unusual-looking double peak in the distance, like two pyramids shoved next to one another, should be useful to someone navigating the arms of the maze. But to a subject who is following instructions, it's irrelevant. They have checked their critical faculties at the starting gate—switched them off. It's as if they're following a recipe so closely that they don't notice a disaster has taken place in the mixing bowl.

This is an important point. Navigation is provisional. It involves the constant monitoring of one's current position, spatial updating, revision, backtracking if needed, the endless consideration and evaluation of numerous different options, and the recalculation of future impending decisions based on each option we take. It's a tightrope act. But following a series of instructions is not the same as navigating. If we rely on a chain of instructions instead—*take the next left turn . . . proceed half a mile . . . take a right turn followed by an immediate left turn*—we're not constructing a cognitive map anymore.

"These types of strategies are not flexible," says Dahmani. "They don't allow you to truly learn how your environment is laid out. If you wanted to take a shortcut you wouldn't be able to because you've never taken that route before and because you can't picture where landmarks are in your mind's eye."

In your mind's eye, she says. With GPS, you have blinded your mind's eye. Who among us hasn't driven home at night with the GPS screen glowing serenely in the blackness, floating like an otherworldly blue rectangle, as if we were on autopilot?

Next, Dahmani asked her subjects to draw a bird's-eye view of the virtual environment they were navigating. "The maps the heavy GPS users drew contained fewer landmarks than the maps of people who used GPS to a lesser extent," she says. In other words, because GPS maps show us fewer landmarks to begin with, the subjects gradually had become trained to notice fewer landmarks in the actual environment. The more they used GPS, the less informative their maps became. "They slowly lose this reflex of looking around," says Dahmani, "even when they're not using GPS and have to find their way by themselves."

GPS technology is ubiquitous. It's everywhere now: in every pocket, from Igloolik to London. We have unleashed it on ourselves. But only now are scientists like Dahmani beginning to ask how GPS might alter the way our brains function, its dark power to erode and corrupt our built-in navigation systems. We didn't even stop to consider it before. Why? Did we assume it was benign? Was it because the brain's spatial workings are mostly hidden from us? Either way, a revolution—or perhaps a devolution—has been taking place in front of us and we haven't noticed. Dahmani's study is one of just a handful of published studies. The effects of GPS remain understudied. But the technology is already everywhere.

"The plan was to recruit people who had a lot of GPS experience and people who didn't have any, or very little, GPS experience," she says. In total, the study took several years to complete. In that shortened time frame, Dahmani noticed a change in our habits: "As time went on, it became harder and harder to find people who didn't have any GPS experience."

•

Hugo Spiers has seen this effect of GPS too. In a 2017 study, the University College London researcher put subjects in an fMRI scanner and monitored their brain activity. Half of his subjects explored virtual London naturally, and the other half followed given instructions instead. Spiers focused his attention on the hippocampus and the prefrontal cortex, a region involved in planning and decision making. These regions become particularly active when navigating environments bursting with different route options. For instance, says Spiers, consider Seven Dials, an intersection in London's West End, near Covent Garden. Even rendered virtually, Seven Dials is a dizzyingly complex environment—seven roads meet in one location like the spokes of an enormous wheel.

"Entering a junction such as Seven Dials in London, where seven streets meet, would enhance activity in the hippocampus, whereas a dead end would drive down its activity," Spiers wrote in a press release that accompanied publication of the study.

As research on sharp-wave ripples has shown, hippocampal neurons are constantly imagining alternative futures, subliminally projecting us into parallel unfurling possibilities. Place cells mentally assess different potential routes and form contingency plans—a Plan B, and a Plan C—in case a chosen route fails.

But not on GPS.

When a subject unquestioningly follows a set of given instructions at Seven Dials, Spiers showed that the hippocampus doesn't perform this act of mental gymnastics. It flatlines. Silence.

•

A general note on cities: they're bad for navigation. They don't sharpen our spatial skills, they dull them. Neuroscientists know this from studying the gargantuan *Sea Hero Quest* dataset. Researcher Antoine Coutrot took a closer look at the scores of around half a million players from thirty-eight different countries, comparing city dwellers with rural subjects. The results were clear. On average, people who grow up in cities perform much worse than people from rural areas.

What's more, not all cities have an equal impact on navigation. For each city, Coutrot calculated its street network entropy (SNE), a measure of what he calls its griddiness. A grid is orderly: the more grid-like a city's layout, the lower its entropy. Entropy is a measure of disorder, so the higher the SNE, the more chaotic and unpredictable the layout of the city. For instance, Chicago is very griddy. With an entropy score of 2.5, it's just about as predictable as a city can be. Consider Prague: medieval, complex, multilayered, organic, haphazard, unpredictable. There isn't a grid in sight. Prague scores 3.6. The difference might sound slight, but it matters to navigation.

Sea Hero Quest players who grew up in high-entropy cities scored better. On average, people raised in Prague beat people from Chicago. Coutrot even took age, gender, and level of education into account to see if they had an effect on scores. It didn't matter. The effect of low-entropy cities on navigation abilities remained. A confusing city—Heraklion in Greece (SNE of 3.54) or Zurich (3.56)—helps us to develop our navigation abilities. Cities built on a grid do the work for us in the same way a smartphone does. American cities are particularly non-challenging: Indianapolis (2.40), New York City (2.53), Philadelphia (2.52), Phoenix (2.10). English cities are bewildering. I was raised in Birmingham (3.56) and went to university in Bristol (3.56), but I am still lost all the time.

•

Meanwhile in Iceland, after driving for more than an hour, Noel Santillan finally pulled his Nissan over to the side of the road. He reentered his destination into his GPS unit. Eight more hours on the road. It couldn't be right. Nonetheless, he kept driving. The GPS had derailed him. Santillan had a serious problem: he was all caudate nucleus now. He was automated. He saw no landmarks. As he steered his car determinedly northward, Santillan's hippocampus, just like those in Spiers's subjects, remained silent. It was flatlining.

When Santillan finally reached Siglufjörður, he stopped in front of the house his GPS had led him to, and knocked on the door. The homeowner informed him that he was off course—hundreds of miles from his intended destination. Tired, he checked into a local hotel.

Then, overnight, something strange happened. As he slumbered, dreaming of icy roads and impossible distances, Santillan became a celebrity. The homeowner had posted an account on social media about the road-tired American who had knocked on her door. While he slept, he had gone viral. Everywhere he went, people recognized him. They shouted his name in the streets as if he was a king returning from battle. During his vacation, he made appearances on radio stations and was interviewed on a TV show.

Being lost didn't bother Santillan. Like a Tsimane child in the Bolivian rainforest, he accepted it—embraced it, even.

•

Scientists have shown what happens to our spatial abilities when we rely too completely on GPS: we stop constructing our own internal maps. There are other ways to stop us from forming cognitive maps. For instance, says Richard Smeyne, send someone to prison. A neuroscientist at Thomas Jefferson University in Philadelphia, Smeyne (an 8 out of 10) studies neurodegenerative diseases like Parkinson's disease. A few years ago, he began studying the brains of prisoners after they had spent long periods isolated in solitary confinement.

"Solitary confinement: we define it as twenty-two to twenty-three hours a day alone in a cell," says Smeyne. "There's often one to two hours out-of-cell, and, in most cases, in that one to two hours they're still isolated. It's just that they're not in the same eight-by-ten space. They may be in another eight-by-ten space. Or it might be a little bit bigger. But they generally still have no meaningful interactions with other people. So that's what we talk about in terms of what we're studying."

Complete isolation. An experiential desert. A tundra of social connections. By any measure, solitary confinement is an extreme environment. Smeyne wanted to see what it would do to the brain. The project began when Smeyne was approached by two lawyers, Jules Lobel at the University of Pittsburgh, and Judy Resnick at Yale University. They were constructing a legal argument: that long-term solitary confinement goes against the Eighth Amendment of the United States Constitution, which prohibits cruel and unusual punishment. Solitary confinement is torture, they said. The evidence is biological.

"People who are in long-term solitary confinement obviously have higher levels of stress," says Smeyne. "They get sick more often. Psychologically, they become very depressed. A proportion of them become psychotic. The suicide rate is significantly higher in prison for people in solitary confinement than those in the general population."

What's more, long-term prisoners are not the only people who suffer the effects of extreme isolation. The elderly, the disabled, and the chronically sick are often isolated too, but in different kinds of prison—in nursing homes and assisted-living facilities, in their malfunctioning bodies, and in hospices.

"Then, there's a category of people who are said to be chronically lonely," says Smeyne. "That's an interesting psychological phenomenon, where people actually are socially integrated but manifest all the symptoms of being alone."

So, Smeyne—bespectacled and sixty-ish, with a neat cap of grey hair—went to prison. He interviewed prisoners who had spent long periods of time in confinement. In some cases, they had spent decades in cramped, brightly lit solitude. He'd asked his subjects what effects long-term isolation had wrought on them during their time in prison, and what effects had lingered after their release. Mostly, what the prisoners told Smeyne didn't surprise him. They suffer depression. They're anxious. They're uncomfortable in large crowds. But there's another consistent and lingering effect of confinement that he hadn't expected: the ex-prisoners are lost.

They no longer know where they are.

"When you think about it neurobiologically, the place cells are our internal GPS and allow us to know the world," he says. "But much of the brain is use-it-or-lose-it. These are the words of Robert King, who was one of the Angola Three and spent thirty-three years in solitary confinement: he said, *I could touch my entire world by extending my arms*."

Convicted of murder in 1974, the Angola Three—King, Albert Woodfox, and Herman Wallace—all spent decades in isolation at the Louisiana State Penitentiary in Angola. King could touch each side of his cell just by turning around, and could walk about ten paces before meeting a wall. Everything was painted the same color. There were no real cues except for where the furniture was bolted to the ground.

"He had no *need* for navigation," says Smeyne. "Even when removed from his cell, he was completely shackled and moved, so again he didn't have to determine where he was going. When he was released from prison—and it's now been, probably, fifteen years since he's been out of prison—he still cannot navigate."

It's almost impossible to study the neurobiological effects of solitary confinement in humans. Researchers can't wheel imaging equipment into prisons, and prisoners can't leave long-term confinement to be scanned. No one has looked at convicts' brains postmortem to assess the effect of living for decades on an emotional tundra. Instead, Smeyne is turning mice into prisoners. From birth to adulthood, the mice are raised in a complex environment, he says. There are multiple generations of mice living together—socializing, fighting over toys and mazes and running wheels. Then comes the tundra.

"At three months we take them and we put them into single caging," he says. One animal to a cage. "We try to mimic the effects of solitary in that they can see other mice, they can talk to other mice, you know, vocalize with other mice, they can smell other mice," but there's no interaction. They are isolated. In solitary confinement, prisoners are relentlessly illuminated—bathed in the fluorescent light that fills every cell. In a superlative example of irony, Smeyne was not allowed to put mice under the same lighting conditions in the lab. The committee that approves Smeyne's animal studies determined that it would be cruel. "It is actually so much harder to put a mouse into isolation than a human being," he says, "which seems inherently wrong."

After either thirty, sixty, or ninety days in solitary confinement, Smeyne measures the effects of confinement on their brains. "We're measuring neuronal volume," he says, "we're measuring neuronal branching, spine density, total dendrite length and total axon length."

The effects of solitary confinement are unsettling. The brain shrinks. And quickly. After just thirty days in isolation, neurons in the hippocampus—the brain structure that contains almost all of the brain's place cells—are 25 percent smaller in volume. The dendrites—the branched extensions that resemble the canopy of a tree and supply information to a neuron—are reduced in length by about 25 percent. It's as if the neuronal branches have been pruned, cut back by a heavy-handed microscopic arborist. The same is true for the axons, the long process by which information travels from the cell body to other neurons. In a normal and healthy neuron, each dendrite is covered with small protrusions known as dendritic spines—the points of contact and communication between the neuron and the thousands of other neurons near it. But in a mouse that has spent time in solitary confinement, there are fewer spines.

By every measure Smeyne used, the hippocampal neurons were impacted by isolation and confinement. Other brain regions like the motor cortex were affected too. The effects were still there after ninety days of isolation. Surprisingly, isolation in female mice had an opposite effect: Smeyne saw neuronal growth instead. He doesn't really understand why. Along with his student Vibol Heng, Smeyne is continuing his research to determine whether the neurobiological effects of solitary confinement are reversible or permanent in mice and humans, and whether women are somehow more resilient to isolation, or even benefit from it.

By some estimates, around 80,000 people are held in some form of solitary confinement in the United States alone. Most of them are men.

"What if I told you that when you come out after long-term solitary, we're only going to take off twenty-five percent of your arm?" says Smeyne. "People would say, *You can't do that. That's torture. You can't take twenty-five percent of the function away from an animal.*" Even so, he says, the changes taking place in the brain are similar. By shrinking in size and losing their points of contact with other neurons, the cells are losing their ability to function.

"Whether it's an arm or a neuron, you're still causing this physical change," says Smeyne. "If we find that we can't return the function and they don't grow back and make connections again, or if we find behavioral changes that aren't reversed, that's when the courts will take this and say, *Okay, you've caused a physical change in the brain that leads to a behavioral change that mimics what you see in people*, and maybe they'll accept it."

Robert King was released in 2001. But once you've spent almost thirty years in solitary confinement, a part of you stays in solitary confinement forever. Smeyne's work on mice suggests King's place cells and hippocampus remained in solitary long after the rest of his body left it behind. It's the reason King no longer knew where he was after he left prison.

•

As a society, despite the occasional campaign promises of politicians, we have failed to redesign prisons. But what if we could redesign GPS? Is it already too late? What if there's a way to stop people from driving blindly across beaches and into the waves, and from driving into the air from unfinished bridges to collide with gravity?

In Berlin, Klaus Gramann thinks there might be. Gramann (whose spatial skills are what he calls a mediocre 5 or 6) is an experimental psychologist. He studies a field known as embodied cognition—which is based on the underlying notion that the brain is inseparable from the body, and that cognition is taking place *within* the body.

A few years ago, Gramann ran an experiment at the Berlin Institute of Technology in which he subtly altered the way his human subjects interface with a GPS unit. Seated in a driving simulator—the sort of thing you might find in a gaming arcade—his subjects took a route through a virtual city, following GPS instructions they were given. A third of the subjects received the kind of instructions we're all familiar with. For instance:

Please turn right at the next intersection.

As we know from the work of Spiers, Dahmani, and others, this kind of instruction deactivates the hippocampus—the industrious source of the mental map.

Another group heard different instructions. At important decision points along the route, they were given instructions that were more descriptive for a particular landmark at the intersection. The instructions had context and meaning. For instance:

Please turn right at the bookstore. Here you can buy books.

A third and final group of subjects received another set of instructions that included some personally relevant modifiers. For instance:

Please turn right at the bookstore. Here you can buy your favorite book, Moby-Dick.

Before sitting them at the simulator, Gramann had collected personal information from each subject about their favorite hobbies, books, movies, and so on. Subjects who received the modified instructions performed better than those who just followed GPS directions. Suddenly, with just these minor alterations, navigating by GPS didn't have such a harmful impact on spatial memory. The subjects had become better at recognizing landmarks.

This could be our future, says Gramann. "If you have all your social media on the cell phone that you use for Google Maps, you have your Friends list, you have your search history online: the system basically knows what you're interested in," he says. Imagine a future in which we give more power to technology, and let our smartphones sift through the data to generate directions that are meaningful to us in a particular and specific way. "Why wouldn't you pull that kind of information out of the system automatically?" says Gramann. "If you could do that in a secure fashion, you could basically provide information in any environment, arbitrarily picking out buildings and aspects of the environment that could relay information based on personal interest."

Please turn right at the coffeeshop where you met your wife for the first time and fell in love.

In her recent study, Louisa Dahmani saw that her subjects don't use landmarks anymore—almost as if they're no longer receptive to them. But Gramann's subjects are. He says that when your navigation system creates artificial landmarks in the environment, populating it with personal meaning, you acquire real landmark knowledge. With landmark knowledge comes route knowledge. Points on the map could be the building you work in; the hospital your son was born in; the dental office where your wisdom teeth were removed; and your favorite Indian restaurant. "Eventually, based on these different landmark configurations, you will acquire a cognitive map-like knowledge—survey knowledge," he says. "And that would trigger neural dynamics in the hippocampal formation, thus training the very system that is necessary for general memory processes."

Gramann's subjects remembered landmarks on the route they took. But they even remembered landmarks better that hadn't been mentioned by the GPS device. By giving subjects context and relevant information, the modified GPS instructions engage their brains in ways that a more traditional GPS device doesn't. Later, when Gramann asks his subjects to sit in the simulator and drive the route through the virtual city again, but this time without GPS instructions, the subjects who received the modified instructions make fewer mistakes.

Since that first study, Gramann has sent subjects into the streets of Charlottenburg, an upscale district in central Berlin, wearing elastic caps that bristle with sixty-five electrodes. They walk along the Kurfürstendamm, past trattorias and cosmetics stores, while he monitors their brain activity. By taking electroencephalogram (EEG) recordings from different brain regions as subjects navigate real-world environments, he believes he'll collect more accurate data than by imaging subjects in an fMRI scanner in a lab setting. After all, navigation happens in the real world, says Gramann.*

But what do we do with findings like Gramann's? We could do a lot. His studies show that a few subtle changes to the GPS devices in every car and smartphone could make us safer, reducing the number of deaths by GPS.

When I spoke with Sharon Roseman, one of Giuseppe Iaria's topographical disorientation subjects, she told me that Google computer engineers had visited her at her home in Denver in July 2019. They had come from Googleplex, Google's corporate headquarters in Mountain View, California. After interviewing Sharon, they watched as she sat at her computer and tried to use Google Maps. They wanted to try to make Google Maps easier for her to use.

Perhaps they should actually be trying to make maps harder to use. When someone experiences a loss of brain function after an injury or a sudden stroke, we try to rehabilitate them. Google should be making it more difficult to drive into the ocean or onto a railway track. That much is obvious. But they should also construct a map that forces our place cells to engage with it instead of silencing them. They should make

* Since we get spatial information from our vestibular system and our muscles, and use it to monitor and update our own movements, the information we obtain from subjects lying prone in a scanner is informative, but only to a point.

maps that allow us to form our own mental representations of space. Until then, people will continue to set out for a train station in Brussels and arrive in Zagreb instead. The question is: will Google do it?

•

Either way, we can make our own choices. We should remain fully alive to the idea that the first time we enter a city, we have no mental map for it. At the exact moment that we're at our most lost—maximally disoriented—we're also busiest constructing the tools that will help us find our way. There are ways to make that process more effective.

"I was never a very good navigator," says Eva-Maria Griesbauer, a graduate student working with Hugo Spiers at University College London. In the lab, Griesbauer scans taxi drivers to better understand how they plan their complex routes through the city. Knowing she was a poor navigator, she says she set herself a challenge. Griesbauer decided: "If I look up something on Google Maps, it will only be to understand where the goal location is," she says. "But the planning—the actual planning of how to get there—I will do that."

Very quickly, she began to see positive results—and they were significant. She started to form her own maps. The city revealed itself in new and informative ways.

"I learned what London is like," she says. Soon enough, she was able to plan her own routes through the city and understand where different parts of London are located in relation to one another. Even now, in new and unfamiliar cities, Griesbauer says her modified approach to navigation has paid off.

"I find things a lot quicker in comparison to other people," she says. "I've trained my brain to understand the environment in a different way." Most importantly, she says, this is something we can all choose to do.

"I think a lot of people could do it if they give it a go and use their brains."

Speaking to Griesbauer reminded me of the last time I visited Washington, DC. Early in the morning, I left my hotel on Capitol Hill. In the near distance, at the end of the street, the United States Capitol rose into the quiet morning air like an enormous upside-down ice cream cone. I set off in the other direction with my cell phone in my hand, surrounded by recognizable landmarks, to find a coffee shop. I got lost anyway. Bewildered by the map, I made a wrong turn and then another. I walked around a building. As I walked, I watched the blue cursor on the screen of my cell phone drifting down the street sideways, suddenly defying the physical universe. The dark magic of the city had taken over again. The different brain structures that should be working together to help me navigate had been silenced. Like Dahmani's subjects, I hadn't noticed the landmarks around me—because we don't anymore—and so I hadn't used them to build myself an inner map. It began to rain—lightly at first, and then less lightly. Slate-grey clouds gathered. I drank my coffee.

Outside Union Station, a one-legged man was arguing with himself, pulling a wet sock onto his hand. I interrupted him and asked how to get back to my hotel. The rain fell on us. He swiveled mid-argument and pointed across buildings and streets, toward distant city blocks in city shadow. His cognitive map was fully functional, buzzing with data, constantly updating, packed with information.

I looked at the sock dangling on the hand, turned to see where it pointed, and I began to walk.

Chapter Ten

What Happened to Amanda Eller

On May 24, 2019, Amanda Eller marked her sixteenth morning in the forest. It was a quiet morning, like the others. The sun came up. She was losing weight. Her lips were blistered and cracked. She had a skin infection. Her leg had been fractured in a fall as she tried to descend a waterfall. She moved slowly now. She was crawling. A flash flood had roared down the slopes and taken her shoes.

By then, Eller had begun to know everything about the way the light fell through the trees. She had left the world of appointments and obligations far behind, back in the parking lot with her cell phone and water bottle, and the car key hidden beneath the tire. Every day since her disappearance, volunteers had combed the forested slopes of Mount Haleakalā, calling her name. As they walked, they probed the undergrowth with sticks, checking for something solid hidden in the shadows on the ground—something solid like her. They'd picked through the intestines of wild boar for a physical sign of her. They'd flown drones above the forest canopy. At times, Eller had heard search helicopters flying back and forth overhead. But pilots hadn't seen her waving at them through the branches. She had stayed close to water. A few times, she'd waded into streams and rivers, crouching over the dark water to try to catch crayfish. Failing, she'd survived instead on wild raspberries and moths and whatever else she could find. It rained often. At night, she covered herself in ferns.

By the morning of May 24, Eller had traveled—walked, fallen, and crawled—around seven miles from the parking lot and the makeshift search headquarters. The search efforts had concentrated mostly on a 1.5-mile-wide radius surrounding the parking lot at the trailhead. Lost people don't usually get too far. But she was miles away to the northwest—miles across ravines and wild and tangled country. One mile farther northwest and she'd have hit the Oahu coastline. Searching farther from the usual search zones was an act of desperation by rescuers tired of combing the same thickets and ravines and waterfalls over and over again.

But a helicopter crew saw her. She was sitting on a gravel bed at the edge of a river, in a deep hole between two waterfalls. She later told her rescuers that she was about to crawl into the river and let herself be carried over the lip of a seventy-foot-tall waterfall—a plan that would have taken her into a box canyon and ended in certain death. "I was getting so skinny that I was really starting to doubt if I could survive," Eller told the *New York Times* after her rescue. In the following weeks, from her hospital bed, she was the subject of scathing criticism. Critics were infuriated by statements like this one, which she gave to the *New York Times* after her rescue: "I wanted to go back the way I'd come, but my gut was leading me another way—and I have a very strong gut instinct."

Her rescuers placed her in a basket attached to the helicopter and she was lifted out. She flew up, past branches dancing wildly in the helicopter's downdraft, out of the forest and back into the world. In a single moment, her possible futures suddenly expanded and multiplied again. If you set up the same scenario and run it over and over, there are many versions of the story that don't end this way, with Eller stepping back out of the green hole she had suddenly disappeared into seventeen days earlier.

Remarkably, though, this version did.

•

Not long ago, camping in remote northern Michigan, I took a step off the trail in a cold rain, and I thought again of Amanda. I looked up into a thicket of mossy branches and tried to locate the sunlight that was straining to appear from behind low grey clouds. I searched for a distinctive tree to use as a landmark. Standing beside a tall half-dead pine tree, I asked my place cells to form a map—to do *something*. I petitioned them. I tried to force 10,000 neurons to fire simultaneously and record its location for me. I thought about my head-direction cells, and grid cells, my inner ear, and the different cortical brain regions that decode the forest and everything in it. I thought about what it means to be a modern human versus an early human, or a rat, or a bat, or a dung beetle. Or a bird. I thought about egocentric and allocentric frames of reference, and whether I was traveling cross-hill or uphill or downhill into the trees across the muddy ground. I wondered if—just like it does for J.N. and other people with DTD—my retrosplenial cortex was failing, in some fundamental way, to aggregate the spatial information about my environment.

Perhaps, I tell myself, I'm closer now than ever to being able to form some kind of informative mental map. The next time I visit Chicago, I plan to subject myself once again to the mind-bending realities of the mirror maze.

Most of all, Eller reminds me that getting lost in dense woodland is not an amusing anecdote. It's not a punchline or a harmless eccentricity. It can have terrible consequences. It can be deadly. Even so, perhaps I understand it better now. In images taken for the Human Connectome Project, the brain is rendered as tangled, multicolored bundles of nerve fibers, like bright skeins of yarn—millions of connections corkscrewing through grey matter, crossing and intertwining with one another. It's a supercomputer with a hundred billion connections. My spatial problems reside somewhere in that tangle. My wife's impressive abilities are explained by the particulars of her supercomputer too.

Navigation is one of the most complex cognitive tasks that we perform, and we're doing it constantly—a thousand times a day. Now that I understand it better, it only seems more complex than it did before. In the next decade, scientists like Giuseppe Iaria might find the specific connections—or misconnections—that lead to disorders like topographical disorientation in his subjects, and in people like me. He might identify the precise neuroanatomical root of the spatial failures experienced by people like Patient One, and by entire families that are unfortunate enough to retain them and pass them on to future generations.

In her lab at University of California, Irvine, Elizabeth Chrastil might determine the way differences between my white matter and my wife's impact my ability to walk in a circle. Russell Epstein might discover yet another cortical region—a postage-stamp-sized area on the surface of the brain—that helps us to decode the physical properties of an environment and stops us from walking into a wardrobe when we're trying to leave a room. Perhaps none of them will find anything at all. The brain gives up its secrets reluctantly.

But, occasionally, discoveries still get made. Hugo Spiers has found that another fundamental part of navigation—the backtracking we have to perform when a pathway is blocked—is associated with a brain region known as the anterior cingulate nucleus.

•

My failure to navigate is part of who I am. It's tangled up with everything else about me. It's a product of my brain structure, aspects of which were inherited, and other parts of which were obtained almost imperceptibly a million different ways—by culture, and habit, and by every chance encounter and new experience.

Occasionally, I email Amanda Eller and ask her what happened out in the forest when she was lost. She answers my emails, but I think her experience will continue to remain a mystery to both of us. She gives her spatial skills a 5 out of 10. When I find the time, I sit at my computer and take one of Iaria's training modules for people with topographical disorientation. I don't know if it helps. I don't think I can feel my hippocampus responding to it—suddenly expanding like a sponge dropped in water. Would I if it did? If I'm walking a new route through my neighborhood, I'll turn every few minutes to view it from a different perspective. Temple University researcher Nora Newcombe tells me it's important to do that: to turn and ask, *What does this intersection look like from another direction?*

Sometimes, when I'm driving by myself, I turn off the navigation system and drive unaided through unfamiliar streets. I disappear. At every decision point, I drift farther and farther off route. As I drive, I think about Edward Tolman in 1945, watching his rats navigate the corridors and dead ends of a maze. At different times, I have both found my way home like this, and also not found my way home like this. Once, I drove for what seemed like a long time, until the sun sank behind houses and I only caught glimpses of it, flashing in alleyways and across sudden backyards and open spaces. That time, eventually, I turned my GPS back on to get home.

Texas A&M University researcher Heather Burte once told me that trying hard to navigate doesn't make us perform any better. That was a gift. She's published studies that prove it. "When you tell poor sense-of-direction people, *Listen, I'm going to test you on the neighborhood you're about to walk through: learn it really well*, they can't learn it any better than if they didn't know that," she says.

Sometimes at night, my thoughts drift toward the global positioning system satellites orbiting the planet. Currently, there are thirty-one of them, forming an operational constellation that provides near-global coverage to lost Earthlings. They zip around the planet like electrons orbiting the central core of an atom. The first one was launched in 1978. Who maintains them? What if they fail? What if, one by one, they go dark and tumble uselessly through space?

This morning, I sit in a bright shaft of sunlight, folding paper. I put a sharp crease in it with my thumbnail, following the instructions to make an origami paper crane. Carefully, I bring the tops of its wings together into two fine points.

I turn it over in my hands, and I try to turn it over in my mind.

Appendix I

What to Do If You're Lost in the Wild

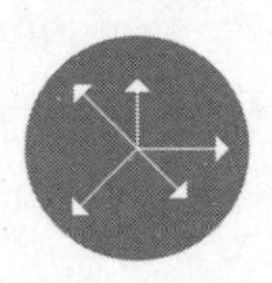

Even in our lostness, we can be predictable, says Robert Koester. In 2008, Koester published *Lost Person Behavior*, a textbook now used worldwide as a field manual by search-and-rescue teams. Unsurprisingly, he rates his own spatial skills at a 9 out of 10. For almost two decades, he's been building an enormous database—a swirling cloud of data. He calls it the International Search and Rescue Incident Database. It contains millions of individual data points from almost 150,000 lost people: disoriented hunters, spelunkers, honeymooners, birdwatchers, mushroom pickers, escaped toddlers, sleep-walkers, moonshiners, survivalists, abductees, Alzheimer's patients who have wandered into their pasts, and lost hikers like Amanda Eller. A 2012 study crunched more than a decade of data from Yosemite National Park and found that the average lost person is a solitary thirty-six-year-old man. He goes missing on a Saturday afternoon in July.

When someone is reported missing, Koester begins collecting data. He'll put a pin in a map to mark an individual's last known location, and then start to scrutinize the local geography. Koester wants to know: When an individual gets lost in the wild, does he follow the river as it drops into the valley, or does he climb uphill instead? Does he cross from one watershed to another? Does he follow a trail? Will a search party find him in a swamp, or on a mountainside, or a boulder field? Is he injured? Suicidal? Will he be found a hundred meters from where he was last seen? When a hiker gets lost in Grand Canyon National Park, does he try to climb out of the canyon in the surging heat, or walk deeper into it? The possibilities are endless.

If two hikers get lost in the same knot of woodland a month apart, they might respond by heading in completely opposite directions. Each subsequent choice—take this fork, follow that path, stay in one place—leads to a multitude of new options, radiating outward forever and terminating in very different places. But when Koester combines the data from enough lost people, patterns begin to emerge. From those patterns, he can calculate probabilities that determine the likeliest locations a lost person might be found.

"I have no clue where the missing person is," says Koester. "All I can tell you is, if 100 people got lost from this location, this is what the distribution would look like. There's going to be more likely areas and there's going to be less likely areas, and you should probably go and look in the more likely areas first."

With careful preparation, the majority of lost person incidents can be avoided in the first place, say Koester and Brett Stoffel, president of Emergency Response International, based in Washington, DC. Most importantly, always tell someone where you are hiking or camping, and when you plan to return. At the least, you should be prepared to make shelter, start a fire, signal for help, disinfect water, and treat minor injuries. Many people get into difficulties when they overestimate the distance they can travel in a single day. Travel in a group: 72 percent of search-and-rescue incidents involve a single person.

Stoffel and Koester say to bring a map and compass and *know how to use them*. Consider using a GPS device, and bring a fully charged cell phone, but don't rely on it. Pay attention to your surroundings and take note of landmarks, especially if you leave the trail. During a hike, turn around often to view your surroundings from different angles and viewpoints. Despite all these precautions, sometimes people become lost. This is what you should do if you're lost in the wild:

1. STOP

Don't act immediately. Instead, take a breath and stay calm. Drink some water and eat a snack. Spatial anxiety is real. People with DTD suffer from it to a greater degree than most people, but we all experience it at one time or another. Think about the last time you knew your location for certain. Look at your map. If you don't have a map, build one. If necessary, draw a map with a stick in the dirt.

2. SIGNAL

Attempt to call 911 on your cell phone. Sometimes, even if a call is not possible, a text message can still be sent. If you have a Personal Locator Beacon, activate it and wait.

3. DECIDE

- **a) Backtrack:** if you can retrace your path safely and accurately to your start point, you should do so.
- **b) Stay put:** if you're hopelessly lost and can't backtrack, and you shared your travel plans with someone before the trip, you will be missed. A search party will come and find you. Stay in one place: don't be a moving target.
- **c) Bail out:** with a map and compass, you should be able to select a direction that will get you to a road or town. Avoid moving in a direction that takes you farther into danger, such as toward impassable cliffs or water that can't be crossed.

4. MOVE

If you can't backtrack, stay put, or bail out, try to get to one of the following, because search-and-rescue teams will search them first:

a) The last place you were seen.

b) Your intended travel route and destination.

c) Known hazards (it sounds counterintuitive, but search-and-rescue teams will often check known hazards).

d) Along linear routes, like roads, trails, railroads, power and utility lines, and water features such as rivers.

In almost all instances—Gerry Largay on the Appalachian Trail, Amanda Eller on Maui, Tyler Batch in the Oregon backwoods—the ordeal of being lost in the wild was preventable.

Don't be a statistic.

Appendix II

How to Become a Better Navigator

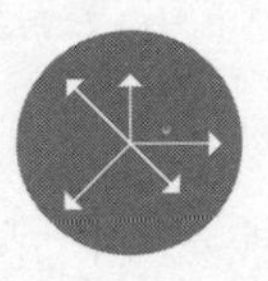

Whether we're lost in the wilderness or simply trying to find our way in an unfamiliar and confusing office building, we can try to improve our spatial skills. Do you want to model your brain on the London cabbie, or the bus driver who endlessly drives the same route? Do you want the spatial awareness of a Tsimane hunter in the Bolivian rainforest, or a GPS-addicted city dweller? The brain is flexible, constantly adapting and rebuilding itself, forming new memories, making connections, recognizing patterns. It forms complex representations of the world. But through habit, we can try to make those mental representations of space even more meaningful.

Temple University cognitive psychologist Nora Newcombe has given me these six simple habits to try to follow as we build our inner maps.

1. Pay attention! Keeping track of where you are and building spatial knowledge is not automatic. It requires conscious effort.
2. When you make a turn, remember that not every turn is a 90-degree angle. Coding a turn as simply a right turn or a left turn is deceptive.
3. As you travel through an environment, look around often. Look behind you to help build an informative cognitive map.
4. Identify a stable landmark and orient to it at all times. Mountains, lakes, oceans, a cell phone tower, a highway. Even if it isn't visible at

every moment, you can begin to keep in mind where it is in relation to your position.

5. Don't just use visual cues. We navigate with all of our senses. Environments bristle with data that we don't want to miss, like sound and smells. In some cities, factors like the slope of the ground can help us to determine a general position.

6. If you are navigating inside a building, look out of every window that you pass in order to maintain your orientation to the external world and its landmarks. Try to determine the footprint of the building to provide a helpful spatial anchor.

Acknowledgments

This book wouldn't have been possible without the help of lots of people. Most importantly, I'd like to thank all the scientists I spoke with, including: Elissa Aminoff, Claudio Aporta, Timothy Behrens, Véronique Bohbot, Neil Burgess, Ariane Burke, Ford Burles, Heather Burte, György Buzsáki, Elizabeth Cashdan, Elizabeth Chrastil, Fred Coolidge, Antoine Coutrot, Jeremy Crampton, Louisa Dahmani, Daniel Dilks, Nico Dosenbach, Paul Dudchenko, Arne Ekstrom, Russell Epstein, André Fenton, Loren Frank, Klaus Gramann, Eva-Maria Griesbauer, Mary Hegarty, Vibol Heng, Ralph Holloway, Giuseppe Iaria, Joshua Jacobs, Kate Jeffery, Robert Koester, Lukas Kunz, Allen Lin, Eleanor Maguire, Edvard Moser, Alysson Muotri, Lynn Nadel, Aurel Nagy, Simon Neubauer, Nora Newcombe, Kaili Rimfeld, Noel Santillan, Richard Smeyne, Hugo Spiers, Larry Squire, Brett Stoffel, Ian Tattersall, Ben Trumble, Jeffrey Taube, Menno Witter, Thomas Wolbers, and Harald Wolf.

Thanks to Breena Kerr for her kind assistance. Thanks to D. Dauphinee for his excellent book on Gerry Largay's disappearance, *When You Find My Body*.

Thanks to all the lost people, especially Amanda Eller, Nadine

Bonnett, Janice Nathan, Sharon Roseman, and Scott Kelbell. And the unlost, like Thierry Gueorgiou.

Thanks to the book people: my agent Katherine Flynn, and my first editor Quynh Do, who saw the value in the project. Endless thanks to editor Melanie Tortoroli, and fixer Mo Crist, and editor Dassi Zeidel, and copyeditor Jodi Beder, who all carried it across the finish line. In the UK, thanks to Cecily Gayford and Ellen Johl at Profile/Wellcome. Thanks in advance to all the bad reviewers: I like your reviews most of all.

Thanks to my buddy Nate Levine for the brain images, and to my good friend Charlie White for images that weren't used but were valuable to me regardless. Thanks to my bosses Caryl and Jack at Michigan State University for providing a great workplace.

Most importantly, I have succeeded thanks to the care and love of my entire family. Where would I be—literally, *where* would I be?—without my spirit guides Max, Izzy, and Rowan? And how well would I fare if I didn't borrow Emeline's impeccable and beautiful brain every day?

Notes

AUTHOR'S NOTE

xi **writing of this book:** Paul A. Dudchenko, *Why People Get Lost* (New York: Oxford University Press, 2010); Arne D. Ekstrom et al., *Human Spatial Navigation* (Princeton, NJ: Princeton University Press, 2018).

CHAPTER ONE: WHERE IS AMANDA ELLER?

2 **parking her SUV at the trailhead:** Breena Kerr, "Amanda Eller, Hiker Lost in Hawaii Forest, Is Found Alive after 17 Days," *New York Times*, May 25, 2019.

2 **But she is lost in a multitude of trees:** Breena Kerr and Alex Horton, "Official Searched 3 Days for a Lost Hiker. Volunteers Wouldn't Quit—and Found Her Weeks Later," *Washington Post*, May 26, 2019.

CHAPTER TWO: PINK SEAHORSES

8 **like eels, or migratory birds:** Connie X. Wang et al., "Transduction of the Geomagnetic Field as Evidenced from alpha-Band Activity in the Human Brain," *eNeuro* 6, no. 2 (Apr. 26, 2019): 0483-18; doi: 10.1523/ENEURO.0483-18.2019.

9 **Arantius carefully hefted a fresh human brain onto his workbench:** Shyamal C. Bir et al., "Julius Caesar Arantius (Giulio Cesare Aranzi, 1530–1589) and the Hippocampus of the Human Brain: History behind the Discovery,"

Journal of Neurosurgery 122, no. 4 (2015): 971–75; doi: 10.3171/2014.11.JNS132402.

9 **survey the brain in the pearly light:** Eliasz Engelhardt, "Hippocampus Discovery: First Steps," *Dementia & Neuropsychologia* 10, no. 1 (2016): 58–62; doi: 10.1590/S1980-57642016DN101 00011.

10 **dark powers to kill three people:** Alan MacFarlane, *Witchcraft in Tudor and Stuart England: A Regional and Comparative Study* (London: Routledge, 1970), xxi.

10 **but it had only just begun:** Andrew Colin Gow, ed., *Witchcraft and the Act of 1604*, Studies in Medieval and Reformation Traditions, vol. 131 (Leiden: Brill, 2008).

10 **the misfortune of a man called H.M.:** Howard Eichenbaum, "What H.M. Taught Us," *Journal of Cognitive Neuroscience* 25, no. 1 (Jan. 2013): 14–21; doi: 10.1162/jocn_a_00285; PMID: 22905817.

12 **". . . and not to the genuine past":** William James, *The Principles of Psychology* (New York: Holt, 1890), 647.

12 **repeat a string of numbers back to researchers:** William Beecher Scoville and Brenda Milner, "Loss of Recent Memory after Bilateral Hippocampal Lesions," *Journal of Neurology, Neurosurgery, and Psychiatry* 20, no. 1 (Feb. 1957): 11–21; doi: 10.1136/jnnp.20.1.11; PMID: 13406589; PMCID: PMC497229.

13 **ten times greater than previous estimates:** Jeneen Interlandi, "New Estimate Boosts the Human Brain's Memory Capacity10-Fold," *Scientific American: Mind*, Feb. 5, 2016, https://www.scientificamerican.com/article/new-estimate-boosts-the-human-brain-s-memory-capacity-10-fold/.

14 **navigation becomes completely impossible:** Howard Eichenbaum, "The Role of the Hippocampus in Navigation Is Memory," *Journal of Neurophysiology* 117, no. 4 (Apr. 1, 2017): 1785–96; doi: 10.1152/jn. 00005.2017; PMID: 28148640; PMCID: PMC5384971.

14 **that leads to swelling in brain tissues:** Oliver Sacks, "The Abyss," *The New Yorker*, Sept. 24, 2007, 100–12; Barbara A. Wilson, Alan D. Baddeley, and Narinder Kapur, "Dense Amnesia in a Professional Musician Following Herpes Simplex Virus Encephalitis," *Journal of Clinical and Experimental Neuropsychology* 17, no. 5 (Oct. 1995): 668–81; doi: 10.1080/01688639508405157; PMID: 8557808.

17 **adrift in a green sea:** Maine Warden Service Report on Gerry's Largay's Disappearance, Nov. 12, 2015.

17 **when the car broke down:** KTVL News 10, "Missing Port Orford Man Located 'Very Hungry' but OK," https://ktvl.com/news/local/sheriff-search-for-missing-port-orford-man.

18 **for the next seven days:** "Man Survives a Week Lost in the Forest," *Curry Coastal Pilot*, Jan. 19, 2018.

18 **leading him to water every day:** Elizabeth Unger, "Lost Tourist Says Monkeys Saved Him in the Amazon," *National Geographic*, Mar. 23, 2017; https://www.nationalgeographic.com/news/2017/03/monkeys-saved-lost-tourist-bolivian-amazon-shamans/.

18 **decided to ransack a hunter's campsite:** Jeff Farrell, "Naked Student Who Got Lost in the Woods for a Month, 'Was on Meth,' Say Police," *The Independent*, Aug. 25, 2017, https://www.independent.co.uk/news/world/americas/naked-student-lost-woods-month-meth-drugs-police-lisa-theris-us-highway-midland-alabama-a7910766.html.

18 **Bullock County Sheriff Raymond Rodgers told the *Daily Mail*:** Ben Ashford, "Woman Who Survived a Month in Woods Was High on Meth," *Daily Mail*, Aug. 22, 2017.

18 **within a hundred yards of her camp:** Kathryn Miles, "The Last Days of Hiker Gerry Largay," *Boston Globe*, Aug. 24, 2016.

19 **geographer Jeremy Crampton (a 6 with a map):** Jeremy W. Crampton, "The Cognitive Processes of Being Lost," *Scientific Journal of Orienteering* 4, no. 1 (1988): 34–36.

20 **he didn't belong to the Communist Party:** Benbow F. Ritchie, "Edward Chace Tolman 1886–1959," *National Academy of Sciences* 1964, http://www.nasonline.org/publications/biographical-memoirs/memoir-pdfs/tolman-edward.pdf.

21 **could track their progress through the maze:** C. James Goodwin, "A-mazing Research," *Monitor on Psychology* 43, no. 2 (2012), https://www.apa.org/monitor/2012/02/research.

21 **radiating outward like the spokes of a wheel:** Edward C. Tolman, "The Determiners of Behavior at a Choice Point," *Psychological Review* 45, no. 1 (1938): 1–41.

21 **the spatial memory of bees:** Elizabeth E. W. Samuelson et al., "Effect of Acute Pesticide Exposure on Bee Spatial Working Memory Using an Analogue of the Radial-Arm Maze," *Scientific Reports* 6 (Dec. 13, 2016); doi: 10.1038/srep38957; PMID: 27958350; PMCID: PMC5154185.

21 **between the clear walls of a saltwater-filled T-maze:** Christelle Alves et al., "Orientation in the Cuttlefish *Sepia officinalis*: Response versus Place Learning," *Animal Cognition* 10, no. 1 (Jan. 2007): 129–36.

22 **echolocating their way:** Theresa M. A. Clarin et al., "Foraging Ecology Predicts Learning Performance in Insectivorous Bats," *PLoS One* 8, no. 6 (June 5, 2013); e64823; doi: 10.1371/journal.pone.0064823.

22 **to other ants in its colony:** A. Ya Karas and G. P. Udalova, "The Behavior of

Ants in a Maze in Response to a Change from Food Motivation to Protective Motivation," *Neuroscience and Behavioral Physiology* 31, no. 4 (July–Aug. 2001): 413–20; doi: 10.1023/a:1010440729177. PMID: 11508492.

22 **then again as an adult moth:** Douglas J. Blackiston, Elena Silva Casey, and Martha R. Weiss, "Retention of Memory through Metamorphosis: Can a Moth Remember What It Learned as a Caterpillar?," *PLoS One* 3, no. 3 (Mar. 5, 2008): e1736; doi: 10.1371/journal.pone. 0001736; PMID: 18320055; PMCID: PMC2248710.

22 **a cell rampaging down a narrow corridor:** Luke Tweedy et al., "Seeing around Corners: Cells Solve Mazes and Respond at a Distance Using Attractant Breakdown," *Science* 369, no. 6507 (Aug. 28, 2020): eaay9792; doi: 10.1126/science.aay9792; PMID: 32855311.

24 **to find her way in an unfamiliar building:** Edward Chace Tolman, "Cognitive Maps in Rats and Men," *Psychological Review* 55 (1948): 189–208.

28 **the caudate nucleus burst into sudden activity:** Giuseppe Iaria et al., "Cognitive Strategies Dependent on the Hippocampus and Caudate Nucleus in Human navigation: Variability and Change with Practice," *Journal of Neuroscience* 23, no. 13 (July 2, 2003): 5945–52, doi: 10.1523/JNEUROSCI.23-13-05945.2003. PMID: 12843299; PMCID: PMC6741255.

30 **by representing them as mathematical models:** Timothy E. J. Behrens et al., "What Is a Cognitive Map? Organizing Knowledge for Flexible Behavior," *Neuron* 100, no. 2 (Oct. 24, 2018): 490–509; doi: 10.1016/j.neuron.2018.10.002; PMID: 30359611.

CHAPTER THREE: IN THE FIRING FIELDS

35 **a neuroscientific, philosophical, and technical manifesto:** John O'Keefe and Lynn Nadel, *The Hippocampus as a Cognitive Map* (New York: Oxford University Press, 1978).

36 **as Nadel put it in a 2014 interview:** Anna Goldenberg, "How Lynn Nadel Helped John O'Keefe Develop Nobel Prize–Winning Research," *Forward*, Oct. 10, 2014.

40 **a rat runs along a narrow, winding, elevated track:** "Hippocampal Place Cells Recorded in the Wilson Lab at MIT," YouTube video, posted by mwlmovies, Oct. 15, 2010, https://www.youtube.com/watch?v=lfNVvoA8QvI.

41 **and help us navigate social dynamics:** Matthew Schafer and Daniela Schiller, "Navigating Social Space," *Neuron* 100, no. 2 (Oct. 24, 2018): 476–89; doi: 10.1016/j.neuron.2018.10.006. PMID: 30359610; PMCID: PMC6226014.

41 **nonspatial information, like sound:** Yoshio Sakurai, "Coding of Auditory

Temporal and Pitch Information by Hippocampal Individual Cells and Cell Assemblies in the Rat," *Neuroscience* 115, no. 4 (2002): 1153–63; doi: 10.1016/s0306-4522(02)00509-2; PMID: 12453487.

41 **or odor:** Howard Eichenbaum et al., "Cue-Sampling and Goal-Approach Correlates of Hippocampal Unit Activity in Rats Performing an Odor-Discrimination Task," *Journal of Neuroscience* 7, no. 3 (Mar. 1987): 716–32; doi: 10.1523/JNEUROSCI.07-03-00716.1987; PMID: 3559709; PMCID: PMC6569079.

41 **Or faces. Or objects:** Itzhal Fried, K. A. MacDonald, and Charles L. Wilson, "Single Neuron Activity in Human Hippocampus and Amygdala during Recognition of Faces and Objects," *Neuron* 18, no. 5 (May 1997): 753–65; doi: 10.1016/s0896-6273(00)80315-3; PMID: 9182800.

43 **in a 1983 paper:** György Buzsáki, L. Stan Leung, and Cornelius H. Vanderwolf, "Cellular Bases of Hippocampal EEG in the Behaving Rat," *Brain Research* 287, no. 2 (1983): 139–71; doi: 10.1016/0165-0173(83)90037-1.

44 **the possible routes a hurricane might take:** Shantanu P. Jadhav et al., "Awake Hippocampal Sharp-Wave Ripples Support Spatial Memory," *Science* 336, no. 6087 (2012): 1454–58; doi: 10.1126/science.1217230.

44 **as removing the entire hippocampus:** Gabrielle Girardeau et al., "Selective Suppression of Hippocampal Ripples Impairs Spatial Memory," *Nature Neuroscience* 12, no. 10 (Oct. 2009): 1222–23; doi: 10.1038/nn.2384; PMID: 19749750.

44 **Without them, the mice are lost:** Lisa Roux et al., "Sharp Wave Ripples during Learning Stabilize the Hippocampal Spatial Map," *Nature Neuroscience* 20, no. 6 (2017): 845–53; doi: 10.1038/nn.4543.

44 **". . . environment it had been in, say, twenty minutes ago":** Mattias P. Karlsson and Loren M. Frank, "Awake Replay of Remote Experiences in the Hippocampus," *Nature Neuroscience* 12, no. 7 (2009): 913–18; doi: 10.1038/nn.2344.

45 **with the *Matrix*-like title, "Constant Sub Second Cycling Between Representations of Possible Futures in the Hippocampus":** Kenneth Kay et al., "Constant Sub-second Cycling between Representations of Possible Futures in the Hippocampus," *Cell* 180, no. 3 (2020): 552–67.e25; doi: 10.1016/j.cell.2020.01.014.

45 **around eight times a second:** Céline Drieu and Michaël Zugaro, "Hippocampal Sequences during Exploration: Mechanisms and Functions," *Frontiers in Cellular Neuroscience* 13, no. 232 (June 13, 2019), doi: 10.3389/fncel.2019.00232.

46 **either was previously in, or hasn't been yet:** Mark C. Zielinski, Wenbo Tang, and Shantanu P. Jadhav, "The Role of Replay and Theta Sequences in Mediat-

ing Hippocampal-Prefrontal Interactions for Memory and Cognition," *Hippocampus* 30, no. 1 (2020): 60–72; doi: 10.1002/hipo.22821.

47 **place cells in epilepsy patients:** Arne D. Ekstrom et al., "Cellular Networks Underlying Human Spatial Navigation," *Nature* 425, no. 6954 (2003): 184–88; doi: 10.1038/nature01964.

48 **an image of a human face:** Nancy Kanwisher et al., "The Fusiform Face Area: A Module in Human Extrastriate Cortex Specialized for Face Perception," *Journal of Neuroscience* 17, no. 11 (1997): 4302–11; doi: 10.1523/JNEUROSCI.17-11-04302.1997.

48 **better at recognizing faces:** Rankin W. McGugin, Ana E. Van Gulick, and Isabel Gauthier, "Cortical Thickness in Fusiform Face Area Predicts Face and Object Recognition Performance," *Journal of Cognitive Neuroscience* 28, no. 2 (2016): 282–94; doi: 10.1162/jocn_a_00891.

48 **electric shocks to the teeth:** Michael Brügger et al., "Tracing Toothache Intensity in the Brain," *Journal of Dental Research* 91, no. 2 (2012): 156–60; doi: 10.1177/0022034511431253.

49 **voxels of unresolved fury:** Thomas F. Denson et al., "The Angry Brain: Neural Correlates of Anger, Angry Rumination, and Aggressive Personality," *Journal of Cognitive Neuroscience* 21, no. 4 (2009): 734–44; doi: 10.1162/jocn.2009.21051.

49 **the dimming of a light:** Hongwen Song et al., "Love-Related Changes in the Brain: A Resting-State Functional Magnetic Resonance Imaging Study," *Frontiers in Human Neuroscience* 9, no. 71 (Feb. 13, 2015), doi: 10.3389/fnhum.2015.00071.

49 **positron emission tomography, or PET:** Eleanor A. Maguire et al., "Knowing Where and Getting There: A Human Navigation Network," *Science* 280, no. 5365 (1998): 921–24. doi: 10.1126/science.280.5365.921.

50 **bus drivers who drive the same streets:** Eleanor A. Maguire, Katherine Woollett, and Hugo J. Spiers, "London Taxi Drivers and Bus Drivers: A Structural MRI and Neuropsychological Analysis," *Hippocampus* 16, no. 12 (2006): 1091–1101; doi: 10.1002/hipo.20233; PMID: 17024677.

51 ***Duke Nukem 3D* for spatial research:** Hugo J. Spiers et al., "Unilateral Temporal Lobectomy Patients Show Lateralized Topographical and Episodic Memory Deficits in a Virtual Town," *Brain* 124, no. 12 (2001): 2476–89; doi: 10.1093/brain/124.12.2476.

54 **the Santa Barbara Sense of Direction Scale:** Mary Hegarty et al., "Development of a Self-Report Measure of Environmental Spatial Ability," *Intelligence* 30 (2002): 425–47.

56 **from millions of people at once:** Antoine Coutrot et al., "Global Determinants of Navigation Ability," *Current Biology* 28, no. 17 (2018): 2861–66.e4;

doi: 10.1016/j.cub.2018.06.009.

57 **carefully reported news stories with headlines:** Tala Salem, "Study: Men Are Better Navigators than Women," *US News and World Report*, May 25, 2018, https://www.usnews.com/news/national-news/articles/2018-05-25/study-men-are-better-navigators-than-women; Fox News, "Men Have a Better Sense of Direction than Women, Study Says," Dec. 8, 2015, https://www.foxnews.com/science/men-have-a-better-sense-of-direction-than-women-study-says.

58 **According to the World Bank's 2018 report "Women, Business and the Law":** World Bank Group, "Women, Business and the Law 2018" (Washington, DC: World Bank. © World Bank, 2018). https://openknowledge.worldbank.org/handle/10986/29498 License: CC BY 3.0 IGO.

58 **had been subject to a decades-old driving ban:** Ben Hubbard, "Saudi Arabia Agrees to Let Women Drive," *New York Times*, Sept. 26, 2017.

CHAPTER FOUR: THE PERCEPTION OF DOORS

60 **his lab at SUNY Downstate Medical Center in Brooklyn:** James B. Ranck Jr., "Foreword: History of the discovery of head direction cells," in *Head Direction Cells and the Neural Mechanisms of Spatial Orientation*, ed. Sidney I. Wiener and Jeffrey S. Taube (Cambridge, MA: MIT Press, 2005).

61 **like finding a new star:** Kashmira Gander, "Endorestiform Nucleus: Scientist Just Discovered a New Part of the Human Brain," *Newsweek.com*, Nov. 22, 2018.

62 **humans also have head-direction cells:** Misun Kim and Eleanor A. Maguire. "Encoding of 3D Head Direction Information in the Human Brain," *Hippocampus* 29 (2019): 619–29; doi: 10.1002/hipo.23060.

62 **when the rat faced east: *pop pop pop pop*:** Jeffrey S. Taube, Robert U. Muller, and James B. Ranck Jr., "Head-Direction Cells Recorded from the Postsubiculum in Freely Moving Rats. II. Effects of Environmental Manipulations," *Journal of Neuroscience* 10, no. 2 (1990): 436–47, doi: 10.1523/JNEUROSCI.10-02-00436.1990.

65 **And neither can a human:** Ryan M. Yoder and Jeffrey S. Taube, "The Vestibular Contribution to the Head Direction Signal and Navigation," *Frontiers in Integrative Neuroscience* 8, no. 32 (Apr. 22, 2014), doi: 10.3389/fnint.2014.00032.

67 **a neurobiologist at the University of Stirling in Scotland:** Paul A. Dudchenko, Emma R. Wood, and Anna Smith, "A New Perspective on the Head Direction Cell System and Spatial Behavior," *Neuroscience and Biobehavioral Reviews* 105 (2019): 24–33; doi: 10.1016/j.neubiorev.2019.06.036.

67 **a discrete region on the surface of the brain:** Seralynne D. Vann, John

P. Aggleton, and Eleanor Maguire, "What Does the Retrosplenial Cortex Do?," Nature Reviews, *Neuroscience* 10, no. 11 (2009): 792–802; doi: 10.1038/nrn2733.

67 **". . . walking from one room to another":** says Dudchenko: Pierre-Yves Jacob et al., "An Independent, Landmark-dominated Head-Direction Signal in Dysgranular Retrosplenial Cortex," *Nature Neuroscience* 20, no. 2 (2017): 173–75; doi: 10.1038/nn.4465.

67 **But not the cells Jeffery was recording:** Hector J. I. Page and Kate J. Jeffery, "Landmark-Based Updating of the Head Direction System by Retrosplenial Cortex: A Computational Model," *Frontiers in Cellular Neuroscience* 12, no. 191 (July 13, 2018), doi: 10.3389/fncel.2018.00191.

70 **they had found the grid:** Torkel Hafting et al., "Microstructure of a spatial map in the entorhinal cortex," *Nature* 436, no. 7052 (2005): 801–6; doi: 10.1038/nature03721.

72 **activity in the brains of epileptic patients:** Joshua Jacobs et al., "Direct Recordings of Grid-like Neuronal Activity in Human Spatial Navigation," *Nature Neuroscience* 16, no. 9 (2013): 1188–90; doi: 10.1038/nn.3466.

72 **". . . you can measure grid cells by fMRI":** Christian F. Doeller, Caswell Barry, and Neil Burgess, "Evidence for Grid Cells in a Human Memory Network," *Nature* 463, no. 7281 (2010): 657–61; doi: 10.1038/nature08704.

73 **the *parahippocampal place area*, or PPA:** Russell Epstein and Nancy Kanwisher, "A Cortical Representation of the Local Visual Environment," *Nature* 392, no. 6676 (1998): 598–601; doi: 10.1038/33402.

74 **who have been blind since birth:** Thomas Wolbers et al., "Modality-independent Coding of Spatial Layout in the Human Brain," *Current Biology* 21, no. 11 (2011): 984–89; doi: 10.1016/j.cub.2011.04.038.

74 **SUBJECT: I saw it:** Pierre Mégevand et al., "Seeing Scenes: Topographic Visual Hallucinations Evoked by Direct Electrical Stimulation of the Parahippocampal Place Area," *Journal of Neuroscience* 34, no. 16 (2014): 5399–405; doi: 10.1523/JNEUROSCI.5202-13.2014;

75 **where visual information gets processed:** Daniel D. Dilks et al., "The Occipital Place Area Is Causally and Selectively Involved in Scene Perception," *Journal of Neuroscience* 33, no. 4 (2013): 1331–36; doi: 10.1523/JNEUROSCI.4081-12.2013.

75 ***transcranial magnetic stimulation*:** Jean-Pascal Lefaucheur, "Transcranial Magnetic Stimulation," *Handbook of Clinical Neurology* 160 (2019): 559–80; doi: 10.1016/B978-0-444-64032-1.00037-0.

77 **either close-up or from far away:** Andrew S. Persichetti and Daniel D. Dilks, "Perceived Egocentric Distance Sensitivity and Invariance

across Scene-selective Cortex," *Cortex* 77, (2016): 155–63; doi: 10.1016/j.cortex.2016.02.006.

77 **while entombed in the scanner:** Andrew S. Persichetti and Daniel D. Dilks, "Dissociable Neural Systems for Recognizing Places and Navigating through Them," *Journal of Neuroscience* 38, no. 48 (2018): 10295–10304; doi: 10.1523/JNEUROSCI.1200-18.2018.

78 **photographs of different indoor environments:** Michael F. Bonner and Russell A. Epstein, "Coding of Navigational Affordances in the Human Visual System," *Proceedings of the National Academy of Sciences of the United States of America* 114, no. 18 (2017): 4793–98; doi: 10.1073/pnas.1618228114.

79 **a London bus, a lighthouse:** Stephen D. Auger, Sinéad L. Mullally, and Eleanor A. Maguire, "Retrosplenial Cortex Codes for Permanent Landmarks," *PloS One* 7, no. 8 (2012): e43620; doi: 10.1371/journal.pone.0043620.

80 **the evening of December 11, 2000:** Tadashi Ino et al., "Directional Disorientation Following Left Retrosplenial Hemorrhage: A Case Report with fMRI Studies," *Cortex* 43, no. 2 (2007): 248–54; doi: 10.1016/s0010-9452(08)70479-9.

81 **". . . from a toilet that was about 20m away":** Ritsuo Hashimoto, Yasufumi Tanaka, and Imaharu Nakano, "Heading Disorientation: A New Test and a Possible Underlying Mechanism," *European Neurology* 63, no. 2 (2010): 87–93; doi: 10.1159/000276398.

82 **the fusiform face area and the parahippocampal place area:** Arthur J. Hudson and Gloria M. Grace, "Misidentification Syndromes Related to Face Specific Area in the Fusiform Gyrus," *Journal of Neurology, Neurosurgery, and Psychiatry* 69, no. 5 (2000): 645–48; doi: 10.1136/jnnp.69.5.645.

83 **far better at navigating through a virtual town:** Louisa Dahmani et al., "An Intrinsic Association between Olfactory Identification and Spatial Memory in Humans," *Nature Communications* 9, no. 1 (Oct. 16, 2018): 4162; doi: 10.1038/s41467-018-06569-4.

83 **cells that code for speed:** Zé Henrique T. D. Góis and Adriano B. L. Tort, "Characterizing Speed Cells in the Rat Hippocampus," *Cell Reports* 25, no. 7 (2018): 1872–1884.e4. doi: 10.1016/j.celrep.2018.10.054.

83 **boundaries in the environment:** Trygve Solstad et al., "Representation of Geometric Borders in the Entorhinal Cortex," *Science* 322, no. 5909 (2008): 1865–68. doi: 10.1126/science.1166466.

83 **combining egocentric and allocentric spatial information:** Patrick A. LaChance, Travis P. Todd, and Jeffrey S. Taube, "A Sense of Space in Postrhinal Cortex," *Science* 365, no. 6449 (2019): eaax4192; doi: 10.1126/science.aax4192.

CHAPTER FIVE: THE OBLIGATE SYMBOLISTS

87 **His nose is wide:** Stephen Wroe et al., "Computer Simulations Show That Neanderthal Facial Morphology Represents Adaptation to Cold and High Energy Demands, but Not Heavy Biting," *Proceedings of the Royal Society of Biological Sciences* 285, no. 1876 (2018): 20180085; doi: 10.1098/rspb.2018.0085.

89 **as far afield as Germany:** Laura T. Buck and Chris B. Stringer. "*Homo heidelbergensis*," *Current Biology* 24, no. 6 (2014): R214–15; doi: 10.1016/j.cub.2013.12.048.

89 **Spain:** Aurélien Mounier, François Marchal, and Silvana Condemi, "Is *Homo heidelbergensis* a Distinct Species? New Insight on the Mauer Mandible," *Journal of Human Evolution* 56, no. 3 (2009): 219–46; doi: 10.1016/j.jhevol.2008.12.006.

89 **Greece:** Günter Bräuer et al., "Virtual Reconstruction and Comparative Analyses of the Middle Pleistocene Apidima 2 Cranium (Greece)," *Anatomical Record* 303, no. 5 (2020): 1374–92; doi: 10.1002/ar.24225.

89 **and France:** Marie-Hélène Moncel et al., "Early Evidence of Acheulean Settlement in Northwestern Europe—la Noira Site, a 700,000 Year-Old Occupation in the Center of France," *PloS One* 8, no. 11 (Nov. 20, 2013): e75529; doi: 10.1371/journal.pone.0075529.

89 **every part of the planet:** Eva K. F. Chan et al., "Human Origins in a Southern African Palaeo-Wetland and First Migrations," *Nature* 575, no. 7781 (2019): 185–89; doi: 10.1038/s41586-019-1714-1.

90 **a collapsed cave in Israel:** Nicholas St. Fleur, "In Cave in Israel, Scientists Find Jawbone Fossil from the Oldest Modern Human Out of Africa," *New York Times*, Jan. 25, 2018.

90 **this time in Greece:** Carl Zimmer, "A Skull Bone Discovered in Greece May Alter the Course of Human Prehistory," *New York Times*, July 10, 2019.

90 **the holes are ancient toothmarks:** Ralph Martins, "Was 'Earliest Musical Instrument' Just a Chewed Up Bone?," *National Geographic*, Mar. 31, 2015.

90 **five perfectly round finger holes:** John Noble Wilford, "Flutes Revised Age Dates the Sound of Music Earlier," *New York Times*, May 29, 2012.

92 **suggests it made all the difference:** Ariane Burke, "Spatial Abilities, Cognition and the Pattern of Neanderthal and Modern Human Dispersals," *Quaternary International* 247 (2012): 230–35; doi: 10.1016/j.quaint.2010.10.029.

94 **a hillside dump near Johannesburg in 1966:** Ralph L. Holloway, "New Australopithecine Dndocast, SK 1585, from Swartkrans, South Africa," *American Journal of Physical Anthropology* 37 (1972): 173–85; doi: 10.1002/ajpa.1330370203.

94 **an estimated 105,000 years old:** Stanislava Eisová, Petr Velemínský, and Emiliano Bruner, "The Neanderthal Endocast from Gánovce (Poprad, Slovak Republic)," *Journal of Anthropological Sciences* 96 (Dec. 31): 139–49; doi: 10.4436/JASS.97005; PMID: 31589589.

94 **the oldest existing brain:** Sonia O'Connor et al., "Exceptional Preservation of a Prehistoric Human Brain from Heslington, Yorkshire, UK," *Journal of Archaeological Science* 38, no. 7 (2011): 1641–54.

94 **a body instantly reduced to ashes:** Pierpaolo Petrone et al., "Heat-Induced Brain Vitrification from the Vesuvius Eruption in C.E. 79," *New England Journal of Medicine* 382, no. 4 (2020): 383–84; doi: 10.1056/NEJMc1909867.

95 **pulling a handkerchief from a closed fist:** Ralph L. Holloway, "On the Making of Endocasts: The New and the Old in Paleoneurology," in *Digital Endocasts,* ed. Emiliano Bruner, Naomichi Ogihara, and Hiroki C. Tanabe, Replacement of Neanderthals by Modern Humans Series (Tokyo: Springer, 2018), 1–8.

95 **obtained with a CT scanner:** Thibaut Bienvenu et al., "Assessing Endocranial Variations in Great Apes and Humans Using 3D data from Virtual endocasts," *American Journal of Physical Anthropology* 145, no. 2 (2011): 231–46; doi: 10.1002/ajpa.21488.

96 **when the human brain gained its characteristic shape:** Simon Neubauer, Jean-Jacques Hublin, and Philipp Gunz, "The Evolution of Modern Human Brain Shape," *Science Advances* 4, no. 1 (Jan. 24, 2018): eaao5961; doi: 10.1126/sciadv.aao5961.

97 **the same shape made by the other hand:** Lauren R. Moo et al., "Interlocking Finger Test: A Bedside Screen for Parietal Lobe Dysfunction," *Journal of Neurology, Neurosurgery, and Psychiatry* 74, no. 4 (2003): 530–32; doi: 10.1136/jnnp.74.4.530.

97 **like understanding numbers:** Manuela Piazza, and Véronique Izard, "How Humans Count: Numerosity and the Parietal Cortex," *Neuroscientist* 15, no. 3 (2009): 261–73; doi: 10.1177/1073858409333073.

97 **[like understanding] time:** John B. Issa et al., "Navigating through Time: A Spatial Navigation Perspective on How the Brain May Encode Time," *Annual Review of Neuroscience* 43 (2020): 73–93; doi: 10.1146/annurev-neuro-101419-011117.

97 **feelings of empathy and forgiveness:** Sabrina Strang et al., "Neural Correlates of Receiving an Apology and Active Forgiveness: An fMRI Study," *PloS One* 9, no. 2 (Feb. 5, 2014): e87654; doi: 10.1371/journal.pone.0087654.

97 **and even happiness:** Wataru Sato et al., "The Structural Neural Substrate of Subjective Happiness," *Scientific Reports* 5, no. 16891 (Nov. 20, 2015), doi: 10.1038/srep16891.

98 **and navigation:** Frederick L. Coolidge, "The Exaptation of the Parietal Lobes in *Homo sapiens*," *Journal of Anthropological Sciences* 92 (2014): 295–98; doi 10.4436/JASS.92013.

98 **the parietal cortex encodes egocentric space:** Andreas Schindler and Andreas Bartels, "Parietal Cortex Codes for Egocentric Space beyond the Field of View," *Current Biology* 23, no. 2 (2013): 177–82; doi: 10.1016/j.cub.2012.11.060.

98 **to visualize his elegant mathematic proofs:** Sandra F. Witelson, Debra L. Kigar, and Thomas Harvey, "The Exceptional Brain of Albert Einstein," *Lancet* 353, no. 9170 (1999): 2149–53; doi: 10.1016/S0140-6736(98)10327-6.

99 ***How to Think Like a Neandertal*:** Thomas Wynn and Frederick L. Coolidge, *How to Think Like a Neandertal* (New York: Oxford University Press, 2013).

99 **as they changed position:** Notger G. Müller et al., "Repetitive Transcranial Magnetic Stimulation Reveals a Causal Role of the Human Precuneus in Spatial Updating," *Scientific Reports* 8, no. 1 (July 5, 2018): 10171; doi: 10.1038/s41598-018-28487-7.

99 **specifically, in the precuneus:** Kyoki Suzuki et al., "Pure Topographical Disorientation Related to Dysfunction of the Viewpoint Dependent Visual System," *Cortex: A Journal Devoted to the Study of the Nervous System and Behavior* 34, no. 4 (1998): 589–99; doi: 10.1016/s0010-9452(08)70516-1.

100 **the bone found in modern humans:** Baruch Arensburg and Anne Marie Tillier, "Speech and the Neanderthals," *Endeavour* 15, no. 1 (1991): 26–28; doi: 10.1016/0160-9327(91)90084-0.

100 **a capacity for speech:** Ruggero D'Anastasio et al., "Micro-biomechanics of the Kebara 2 Hyoid and Its Implications for Speech in Neanderthals," *PloS One* 8, no. 12 (Dec. 18, 2013): e82261, doi: 10.1371/journal.pone.0082261.

100 **a protein involved in language acquisition:** Simon E. Fisher et al., "Localisation of a Gene Implicated in a Severe Speech and Language Disorder," *Nature Genetics* 18, no. 2 (1998): 168–70; doi: 10.1038/ng0298-168.

100 **Neanderthals had the gene too:** Johannes Krause et al., "The Derived FOXP2 Variant of Modern Humans Was Shared with Neandertals," *Current Biology* 17, no. 21 (2007): 1908–12; doi: 10.1016/j.cub.2007.10.008.

100 **"The Neanderthals surprised us a bit there":** Kerri Smith, "Modern Speech Gene Found in Neanderthals," *Nature News*, Oct. 18, 2007, doi: 10.1038/news.2007.177.

100 **". . . it's the subjunctive":** Thomas Wynn and Frederick L. Coolidge, "The Expert Neandertal Mind," *Journal of Human Evolution* 46, no. 4 (2004): 467–87; doi: 10.1016/j.jhevol.2004.01.005.

102 **microcephaly in developing brains:** Fernanda R. Cugola et al., "The Brazil-

ian Zika Virus Strain Causes Birth Defects in Experimental Models," *Nature* 534, no. 7606 (2016): 267–71; doi: 10.1038/nature18296.

103 **more than 200 genes:** Martin Kuhlwilm and Cedric Boeckx, "A Catalog of Single Nucleotide Changes Distinguishing Modern Humans from Archaic Hominins," *Scientific Reports* 9, no. 1 (June 11, 2019): 8463; doi: 10.1038/s41598-019-44877-x.

104 **our beginnings as a species:** Irwin Silverman, Jean Choi, and Michael Peters, "The Hunter-Gatherer Theory of Sex Differences in Spatial Abilities: Data from 40 Countries," *Archives of Sexual Behavior* 36, no. 2 (2007): 261–68; doi: 10.1007/s10508-006-9168-6.

104 ***dual-solution paradigm:*** Alexander Boone, Xinyi Gong, and Mary Hegarty, "Sex Differences in Navigation Strategy and Efficiency," *Memory & Cognition* 46, no. 6 (2018): 909–22; doi: 10.3758/s13421-018-0811-y.

105 **men pause less, travel farther:** Ascher K. Munion et al., "Gender Differences in Spatial Navigation: Characterizing Wayfinding Behaviors," *Psychonomic Bulletin & Review* 26, no. 6 (2019): 1933–40; doi: 10.3758/s13423-019-01659-w.

105 **other differences too:** Alina Nazareth et al., "A Meta-analysis of Sex Differences in Human Navigation Skills," *Psychonomic Bulletin & Review* 26, no. 5 (2019): 1503–28; doi: 10.3758/s13423-019-01633-6.

105 **Women navigate using landmarks:** Albert Postma et al., "Losing Your Car in the Parking Lot: Spatial Memory in the Real World," *Applied Cognitive Psychology* 26 (2012): 680–86; doi: 10.1002/acp.2844.

105 **tasks like mental rotation:** Tim Koscik et al., "Sex Differences in Parietal Lobe Morphology: Relationship to Mental Rotation Performance," *Brain and Cognition* 69, no. 3 (2009): 451–49; doi: 10.1016/j.bandc.2008.09.004.

106 **and spare house keys:** Daniel Voyer et al., "Gender Differences in Object Location Memory: A Meta-analysis," *Psychonomic Bulletin & Review* 14, no. 1 (2007): 23–38; doi: 10.3758/bf03194024.

106 **women played an action video game:** Jing Feng, Ian Spence, and Jay Pratt, "Playing an Action Video Game Reduces Gender Differences in Spatial Cognition," *Psychological Science* 18, no. 10 (2007): 850–55; doi: 10.1111/j.1467-9280.2007.01990.x.

107 **They had closed the gap:** Norbert Jaušovec and Ksenija Jaušovec, "Sex Differences in Mental Rotation and Cortical Activation Patterns: Can Training Change Them?," *Intelligence* 40, no. 2 (2012): 151–62; doi: 10.1016/j.intell.2012.01.005.

107 **people from fifty-three countries responded:** Stian Reimers, "The BBC Internet Study: General Methodology," *Archives of Sexual Behavior* 36 (2007): 147–61; doi: 10.1007/s10508-006-9143-2.

107 **But Norwegian men still outperform them:** Richard A. Lippa, Marcia L. Collaer, and Michael Peters, "Sex Differences in Mental Rotation and Line Angle Judgments Are Positively Associated with Gender Equality and Economic Development across 53 Nations," *Archives of Sexual Behavior* 39, no. 4 (2010): 990–97; doi: 10.1007/s10508-008-9460-8.

107 **Girls, and later women, never catch up:** Jillian E. Lauer, J. E. Eukyung Yhang, and Stella F. Lourenco, "The Development of Gender Differences in Spatial Reasoning: A Meta-analytic Review," *Psychological Bulletin* 145, no. 6: 537–65; doi: 10.1037/bul0000191.

CHAPTER SIX: DEAD RECKONING

110 **the ant finds a dead fly:** Cornelia Buehlmann et al., "Desert Ants Locate Food by Combining High Sensitivity to Food Odors with Extensive Crosswind Runs," *Current Biology* 24, no. 9 (2014): 960–64; doi: 10.1016/j.cub.2014.02.056.

113 **to make sure they can't see:** Harald Wolf, Matthias Wittlinger, and Sarah E. Pfeffer, "Two Distance Memories in Desert Ants—Modes of Interaction," *PloS One* 13, no. 10 (Oct. 10, 2018): e0204664; doi: 10.1371/journal.pone.0204664.

113 **amputated their legs:** Kathrin Steck, Matthias Wittlinger, and Harald Wolf, "Estimation of Homing Distance in Desert Ants, *Cataglyphis fortis*, Remains Unaffected by Disturbance of Walking Behaviour," *Journal of Experimental Biology* 212, no. 18 (2009): 2893–901; doi: 10.1242/jeb.030403.

113 **or through complex mazes:** Tobias Seidl and Rüdiger Wehner, "Walking on Inclines: How Do Desert Ants Monitor Slope and Step Length," *Frontiers in Zoology* 5, no. 8 (June 2, 2008); doi: 10.1186/1742-9994-5-8.

113 **enormous surreal obstacle courses:** Sandra Wohlgemuth, Bernhard Ronacher, and Rüdiger Wehner, "Ant Odometry in the Third Dimension," *Nature* 411, no. 6839 (2001): 795–98; doi: 10.1038/35081069.

113 **the ant is counting its steps:** Matthias Wittlinger, Rüdiger Wehner, and Harald Wolf, "The Desert Ant Odometer: A Stride Integrator That Accounts for Stride Length and Walking Speed," *Journal of Experimental Biology* 210, pt. 2 (2007): 198–207; doi: 10.1242/jeb.02657.

115 **find its retreat at night:** Thomas Nørgaard, Yakir L. Gagnon, and Eric J. Warrant, "Nocturnal Homing: Learning Walks in a Wandering Spider?," *PloS One* 7, no. 11 (2012): e49263; doi: 10.1371/journal.pone.0049263.

115 **to gauge distances:** Mandyam V. Srinivasan, "Going with the Flow: A Brief History of the Study of the Honeybee's Navigational 'Odometer,'" *Journal of Comparative Physiology A, Neuroethology, Sensory, Neural, and Behavioral Physiology* 200, no. 6 (2014): 563–73; doi: 10.1007/s00359-014-0902-6.

115 **an Earth compass:** Tali Kimchi, Ariane S. Etienne, and Joseph Terkel, "A Subterranean Mammal Uses the Magnetic Compass for Path Integration," *Proceedings of the National Academy of Sciences of the United States of America* 101, no. 4 (2004): 1105–9; doi: 10.1073/pnas.0307560100.

115 **it uses dead reckoning:** Gal Aharon, Meshi Sadot, and Yossi Yovel, "Bats Use Path Integration Rather Than Acoustic Flow to Assess Flight Distance along Flyways," *Current Biology* 27, no. 23 (2017): 3650–3657.e3; doi: 10.1016/j.cub.2017.10.012.

116 **information gathered during locomotion:** Valérie V. Séguinot, Jennifer Cattet, and Simon Benhamou, "Path Integration in Dogs," *Animal Behaviour* 55, no. 4 (1998): 787–97; doi: 10.1006/anbe.1997.0662.

116 **In his 1873 letter to *Nature*:** Charles Darwin, "Origin of Certain Instincts," *Nature* 7 (1873): 417–18; doi: 10.1038/007417a0.

121 **as well as sighted people:** Jack M. Loomis et al., "Nonvisual Navigation by Blind and Sighted: Assessment of Path Integration Ability," *Journal of Experimental Psychology General* 122, no. 1 (1993): 73–91; doi: 10.1037//0096-3445.122.1.73.

121 **and can use them to navigate:** Panagiotis Koukourikos and Konstantinos Papadopoulos, "Development of Cognitive Maps by Individuals with Blindness Using a Multisensory Application," *Procedia Computer Science* 67 (2015): 213–22; doi: 10.1016/j.procs.2015.09.265.

122 **the actual volume of the brain regions involved:** Elizabeth R. Chrastil et al., "Individual Differences in Human Path Integration Abilities Correlate with Gray Matter Volume in Retrosplenial Cortex, Hippocampus, and Medial Prefrontal Cortex," *eNeuro* 4, no. 2, ENEURO.0346-16.2017 (Apr. 17, 2017); doi: 10.1523/ENEURO.0346-16.2017.

122 **they're thrown out of the study:** Thomas Wolbers et al., "Differential Recruitment of the Hippocampus, Medial Prefrontal Cortex, and the Human Motion Complex during Path Integration in Humans," *Journal of Neuroscience* 27, no. 35 (2007): 9408–16; doi: 10.1523/JNEUROSCI.2146-07.2007.

123 **the ninety billion or so other neurons:** Frederico A. C. Azevedo et al., "Equal Numbers of Neuronal and Nonneuronal Cells Make the Human Brain an Isometrically Scaled-up Primate Brain," *Journal of Comparative Neurology* 513, no. 5 (2009): 532–41; doi: 10.1002/cne.21974.

125 **a straight line through unfamiliar terrain:** Henry Fountain, "Hiking Around in Circles? Probably, Study Says," *New York Times*, Aug. 20, 2009.

125 **Souman took six subjects to the Bienwald:** Jan L. Souman et al., "Walking Straight into Circles," *Current Biology* 19, no. 18 (2009): 1538–42; doi: 10.1016/j.cub.2009.07.053.

126 **a very specific operating system:** Marie Dacke et al., "How Dung Bee-

tles Steer Straight," *Annual Review of Entomology* 66 (2021): 243–56; doi: 10.1146/annurev-ento-042020-102149.

127 **researchers made little cardboard screens:** Marie Dacke et al., "Dung Beetles Use the Milky Way for Orientation," *Current Biology* 23, no. 4 (Feb. 18, 2013): 298–300; doi: 10.1016/j.cub.2012.12.034; PMID: 23352694.

CHAPTER SEVEN: SOMEWHERE EAST OF TIMBUKTU

130 **He called his first subject Patient One:** Giuseppe Iaria et al., "Developmental Topographical Disorientation: Case One," *Neuropsychologia* 47, no. 1 (2009): 30–40; doi: 10.1016/j.neuropsychologia.2008.08.021.

131 **a seventy-two-year-old architect from Massachusetts:** C. Miller Fisher, "Disorientation for Place," *Archives of Neurology* 39, no. 1 (Jan. 1982): 33–36; doi: 10.1001/archneur.1982.00510130035008. PMID: 7055444.

132 **than their non-concussed teammates:** Liam Heath McFarlane et al., "A Pilot Study Evaluating the Effects of Concussion on the Ability to Form Cognitive Maps for Spatial Orientation in Adolescent Hockey Players," *Brain Injury* 34, no. 8 (2020): 1112–17; doi: 10.1080/02699052.2020.1773537.

134 **structural activity of J.N.'s brain:** Jiye G. Kim et al., "A Neural Basis for Developmental Topographic Disorientation," *Journal of Neuroscience* 35, no. 37, 2015, 12954-69; doi: 10.1523/JNEUROSCI.0640-15.2015.

137 **swing, bounce, balance, slosh, slide:** Alex Mitko and Jason Fischer, "When It All Falls Down: The Relationship between Intuitive Physics and Spatial Cognition," *Cognitive Research* 5, no. 24 (May 19, 2020); doi: 10.1186/s41235-020-00224-7.

138 **beginning around the age of nine:** Ford Burles et al., "The Emergence of Cognitive Maps for Spatial Navigation in 7- to 10-Year-Old Children," *Child Development* 91, no. 3 (2020): e733–e744; doi: 10.1111/cdev.13285.

138 **decreased functional connectivity:** Giuseppe Iaria et al., "Developmental Topographical Disorientation and Decreased Hippocampal Functional Connectivity," *Hippocampus* 24, no. 11 (2014): 1364–74; doi: 10.1002/hipo.22317.

138 **Iaria went on to recruit 120 subjects:** Giuseppe Iaria and Jason J. S. Barton, "Developmental Topographical Disorientation: A Newly Discovered Cognitive Disorder," *Experimental Brain Research* 206 (2010): 189–96; doi: 10.1007/s00221-010-2256-9.

138 **patterns in the presence of DTD:** Sarah F. Barclay et al., "Familial Aggregation in Developmental Topographical Disorientation (DTD)," *Cognitive Neuropsychology* 33, no. 7–8 (2016): 388–97; doi: 10.1080/02643294.2016.1262835.

141 **Korsakoff's syndrome:** Erik Oudman et al., "Route Learning in Korsakoff's Syndrome: Residual Acquisition of Spatial Memory Despite Profound

Amnesia," *Journal of Neuropsychology* 10, no. 1 (2016): 90–103; doi: 10.1111/jnp.12058.

141 **Schmahmann's syndrome:** Georgios P. D. Argyropoulos et al., "The Cerebellar Cognitive Affective/Schmahmann Syndrome: A Task Force Paper," *Cerebellum* 19, no. 1 (2020): 102–25; doi: 10.1007/s12311-019-01068-8.

141 **Turner syndrome:** David Hong, Jamie Scaletta Kent, and Shelli Kesler, "Cognitive Profile of Turner Syndrome," *Developmental Disabilities Research Reviews* 15, no. 4 (2009): 270–78; doi: 10.1002/ddrr.79.

142 **preliminary studies with control subjects:** Michael McLaren-Gradinaru et al., "A Novel Training Program to Improve Human Spatial Orientation: Preliminary Findings," *Frontiers in Human Neuroscience* 14, no. 5 (Jan. 24, 2020); doi: 10.3389/fnhum.2020.00005.

CHAPTER EIGHT: YOUR BRAIN, MY BRAIN, HIS BRAIN, HER BRAIN

147 **he realized he needed to study it:** Benjamin C. Trumble et al., "No Sex or Age Difference in Dead-Reckoning Ability among Tsimane Forager-Horticulturalists," *Human Nature* 27, no. 1 (2016): 51–67; doi: 10.1007/s12110-015-9246-3.

149 **And it's the subtle differences:** Jenny Gu and Ryota Kanai, "What Contributes to Individual Differences in Brain Structure?," *Frontiers in Human Neuroscience* 8, no. 262 (Apr. 28, 2014); doi: 10.3389/fnhum.2014.00262.

150 **his own included:** Evan M. Gordon et al., "Precision Functional Mapping of Individual Human Brains," *Neuron* 95, no. 4 (2017): 791–807.e7; doi: 10.1016/j.neuron.2017.07.011.

151 **based solely on their neuroanatomy:** Seyed Abolfazl Valizadeh et al., "Identification of Individual Subjects on the Basis of Their Brain Anatomical Features," *Scientific Reports* 8, no. 5611 (Apr. 4, 2018); doi: 10.1038/s41598-018-23696-6.

152 **working memory and language function:** Alberto Llera et al., "Inter-individual Differences in Human Brain Structure and Morphology Link to Variation in Demographics and Behavior," *eLife* 8 (July 3, 2019): e44443; doi: 10.7554/eLife.44443.

152 **like happiness and emotional well-being:** Yuta Katsumi et al., "Intrinsic Functional Network Contributions to the Relationship between Trait Empathy and Subjective Happiness," *NeuroImage* 227, no. 117650 (2021); doi: 10.1016/j.neuroimage.2020.117650.

152 **subjects with a thicker frontal cortex didn't:** Joaquim Radua, "Frontal Cortical Thickness, Marriage and Life Satisfaction," *Neuroscience* 384 (2018): 417–18; doi: 10.1016/j.neuroscience.2018.05.044.

152 **associated with subjective happiness:** Wataru Sato et al., "Resting-State Neural Activity and Connectivity Associated with Subjective Happiness," *Scientific Reports* 9, no. 12098 (Aug. 20, 2019), doi: 10.1038/s41598-019-48510-9.

153 **What is determined instead by our environment?:** Kaili Rimfeld et al., "Phenotypic and Genetic Evidence for a Unifactorial Structure of Spatial Abilities," *Proceedings of the National Academy of Sciences of the United States of America* 114, no. 10 (2017): 2777–82; doi: 10.1073/pnas.1607883114.

154 **84 percent of the variance in our spatial abilities:** Margherita Malanchini et al., "Evidence for a Unitary Structure of Spatial Cognition beyond General Intelligence," *NPJ Science of Learning* 5, no. 9 (July 2, 2020), doi: 10.1038/s41539-020-0067-8.

154 **". . . variants that are associated with educational attainment":** Kaili Rimfeld et al., "The Stability of Educational Achievement across School Years Is Largely Explained by Genetic Factors," *NPJ Science of Learning* 3, no. 16 (Sept. 4, 2018), doi: 10.1038/s41539-018-0030-0.

155 **watch him compete in Portugal:** "Follow Thierry 1," YouTube video, 2:04, posted by O-Portugal, March 16, 2017, https://www.youtube.com/watch?v=p-Hp-SLhmZ8.

157 **something known as the Mozart effect:** Frances H. Rauscher, Gordon L. Shaw, and Katherine N. Ky, "Music and Spatial Task Performance," *Nature* 14, no. 365 (Oct. 14, 1993): 611; doi: 10.1038/365611a0; PMID: 8413624.

157 **Mozart, or Philip Glass, or silence:** Kenneth M. Steele et al., "Prelude or Requiem for the 'Mozart Effect'?," *Nature* 400, no. 6747 (1999): 827–28; doi: 10.1038/23611.

157 **Mozart, Bach, and silence: still no effect:** Hanyu Lin and Hui Yueh Hsieh, "The Effect of Music on Spatial Ability," in *Internationalization, Design and Global Development. IDGD 2011. Lecture Notes in Computer Science*, vol. 6775, ed. P. L. Patrick Rau (Berlin and Heidelberg: Springer, 2011).

157 **rats performed better in a maze:** Sun-Young Lim and Hiramitsu Suzuki, "Changes in Maze Behavior of Mice Occur after Sufficient Accumulation of Docosahexaenoic Acid in Brain," *Journal of Nutrition* 131, no. 2 (2001): 319–24; doi: 10.1093/jn/131.2.319.

157 **if they heard Mozart as fetuses:** Frances H. Rauscher, Desix Robinson, and Jason Jens, "Improved Maze Learning through Early Music Exposure in Rats," *Neurological Research* 20, no. 5 (1998): 427–32; doi: 10.1080/01616412.1998.11740543.

157 **debunking the Mozart effect:** Kenneth M. Steele, "Do Rats Show a Mozart Effect?," *Music Perception* 21, no. 2 (Dec. 2003): 251–65, doi: 10.1525/mp.2003.21.2.251; Kenneth M. Steele, Karen E. Bass, and Melissa D. Crook, "The Mystery of the Mozart Effect: Failure to Replicate," *Psychological Sci-*

ence 10, no. 4 (1999): 366–69; doi: 10.1111/1467-9280.00169.

157 **"Mozart Effect–Shmozart Effect":** Jakob Pietschnig, Martin Voracek, and Anton K. Formann, "Mozart Effect–Shmozart Effect: A Meta-Analysis," *Intelligence* 28, no. 3 (2010): 314–23; doi: 10.1016/j.intell.2010.03.001.

158 **expert musicians have a larger corpus callosum:** Gottfried Schlaug et al., "Increased Corpus Callosum Size in Musicians," *Neuropsychologia* 33, no. 8 (1995): 1047–55; doi: 10.1016/0028-3932(95)00045-5.

158 **Jugglers show increases in grey matter:** Bogdan Draganski et al., "Neuroplasticity: Changes in Grey Matter Induced by Training," *Nature* 427, no. 6972 (2004): 311–12; doi: 10.1038/427311a.

158 **alters the architecture of the brain:** Sayuri Hayakawa and Viorica Marian, "Consequences of Multilingualism for Neural Architecture," *Behavioral and Brain Functions* 15, no. 6 (Mar. 25, 2019); doi: 10.1186/s12993-019-0157-z.

158 **activated by spatial memory tasks:** Timothy A. Keller and Marcel Adam Just, "Structural and Functional Neuroplasticity in Human Learning of Spatial Routes," *Neuroimage* 125 (Jan. 15, 2016): 256–66; doi: 10.1016/j .neuroimage.2015.10.015; PMID: 26477660.

158 **researchers gave Siberian chipmunks two different substances:** Minghui Wang et al., "Improved Spatial Memory Promotes Scatter Hoarding by Siberian Chipmunks," *Journal of Mammalogy* 99, no. 5 (Oct. 10, 2018): 1189–96; doi: 10.1093/jmammal/gyy109.

159 **the anterior part of it, located toward the front of the brain, becomes smaller:** Katherine Woollett and Eleanor A. Maguire, "Navigational Expertise May Compromise Anterograde Associative Memory," *Neuropsychologia* 47, no. 4 (2009): 1088–95; doi: 10.1016/j.neuropsychologia.2008.12.036.

161 **an uphill, or downhill, or cross-hill component to it:** Peggy Li et al., "Spatial Reasoning in Tenejapan Mayans," *Cognition* 120, no. 1 (2011): 33–53. doi: 10.1016/j.cognition.2011.02.012.

161 **10 percent of all words spoken in a Guugu Yimithirr conversation:** Guy Deutscher, "Does Your Language Shape How You Think?," *New York Times*, Aug. 26, 2010.

162 **they raised mice in total darkness from birth:** Sinisa Hrvatin et al., "Single-Cell Analysis of Experience-Dependent Transcriptomic States in the Mouse Visual Cortex," *Nature Neuroscience* 21, no. 1 (2018): 120–29. doi: 10.1038/s41593-017-0029-5.

164 **animal movements, dreams, and other clues:** Claudio Aporta and Eric Higgs, "Satellite Culture: Global Positioning Systems, Inuit Wayfinding, and the Need for a New Account of Technology," *Current Anthropology* 46, no. 5 (2005): 729–53.

CHAPTER NINE: THE FUTURE

167 **In a 2017 article about Santillan:** *Iceland Magazine*, "American Traveler Who Became Famous for Getting Lost in Iceland Announces His Return," Feb 6, 2017; https://icelandmag.is/tags/noel-santillan.

168 **His wife died:** Peter Holley, "Driver Follows GPS off Demolished Bridge, Killing Wife, Police Say," *Washington Post*, Mar. 31, 2015.

168 **a remote and barely-there road:** Lorna Dueck, "Seven Weeks in Wilderness: Rita Chretien Recalls Her Nightmare," *Globe and Mail*, Oct. 4, 2012.

168 **stranded in mud:** Michelle Lou, "Nearly 100 Drivers Followed a Google Maps Detour—and Ended Up Stuck in an Empty Field," CNN, June 27, 2019, https://edition.cnn.com/2019/06/26/us/google-maps-detour-colorado-trnd/index.html?no-st=1561671662.

168 **they typed Carpi instead:** BBC, "Swedes Miss Capri after GPS Gaffe," July 28, 2009, http://news.bbc.co.uk/2/hi/europe/8173308.stm.

169 **catastrophic GPS-related incidents:** Allen Yilun Lin et al., "Understanding 'Death by GPS': A Systematic Study of Catastrophic Incidents Associated with Personal Navigation Technologies," *Proceedings of the 2017 CHI Conference on Human Factors in Computing Systems*, Association for Computing Machinery, 1154–66; doi: 10.1145/3025453.3025737.

169 **almost all the way to Zagreb:** Alex Davies, "A Woman Drove 900 miles instead of 50 because of Bad GPS Directions," *Business Insider*, Jan. 16, 2013, https://www.businessinsider.com/gps-gives-wrong-directions-to-woman-2013-1.

169 **almost 500 miles from La Plagne:** BBC, "GPS Fail on Bus Sends Belgian Tourists on 1200km Detour," Mar. 10, 2015, https://www.bbc.com/news/world-europe-31814083#:~:text=A%20group%20of%20Belgian%20tourists,to%20France's%20border%20with%20Spain.

170 **standing on top of her submerged car:** Guelph Mercury, "GPS to Ontario Driver: Make Left Turn into Swamp, Wait on Roof for an Hour," Oct. 6, 2010, https://www.guelphmercury.com/news-story/2704021-gps-to-ontario-driver-make-left-turn-into-swamp-wait-on-roof-for-an-hour/.

170 **stranded in snow for three days:** CBC News, "GPS-stranded Driver Stuck in Snow for 3 Days," Mar. 3, 2011, https://www.cbc.ca/news/canada/new-brunswick/gps-stranded-driver-stuck-3-days-in-snow-1.993878.

171 **outside a grocery store in Igloolik:** Claudio Aporta and Eric Higgs, "Satellite Culture: Global Positioning Systems, Inuit Wayfinding, and the Need for a New Account of Technology," *Current Anthropology* 46, no. 5 (2005): 729–53.

173 **effects of GPS on spatial memory:** Louisa Dahmani and Véronique D. Bohbot, "Habitual Use of GPS Negatively Impacts Spatial Memory during Self-guided Navigation," *Scientific Reports* 10, no. 6310 (Apr. 14, 2020), doi: 10.1038/s41598-020-62877-0.

176 **comparing city dwellers with rural subjects:** Antoine Coutrot et al., "Cities Have a Negative Impact on Navigation Ability: Evidence from 38 Countries," *bioRxiv*, Jan. 24, 2020, doi: 10.1101/2020.01.23.917211.

177 **He reentered his destination into his GPS unit:** David Kushner, "Is Your GPS Scrambling Your Brain?," *Outside Magazine*, Nov. 15, 2016.

182 **altered the way his human subjects interface with a GPS unit:** Klaus Gramann, Paul Hoepner, and Katja Karrer-Gauss, "Modified Navigation Instructions for Spatial Navigation Assistance Systems Lead to Incidental Spatial Learning," *Frontiers in Psychology* 8, no. 193 (Feb. 13, 2017), doi: 10.3389/fpsyg.2017.00193.

184 **elastic caps that bristle with sixty-five electrodes:** Anna Wunderlich and Klaus Gramann, "Brain Dynamics of Assisted Pedestrian Navigation in the Real-World," *bioRxiv* 2020.06.08.139469 (June 8, 2020), doi: 10.1101/2020.06.08.139469.

CHAPTER TEN: WHAT HAPPENED TO AMANDA ELLER

188 **radius surrounding the parking lot at the trailhead:** Personal communication with one of Eller's rescuers, Javier Cantellops, Sept. 2020.

188 **ended in certain death:** Naia Carlos, "This Is How Missing Hiker Amanda Eller Survived Two Weeks in Hawaii Forest," *Tech Times*, May 27, 2019, https://www.techtimes.com/articles/243783/20190527/this-is-how-missing-hiker-amanda-elle-survived-two-weeks-in-hawaii-forest.htm.

188 **Eller told the *New York Times* after her rescue:** Breena Kerr, "Amanda Eller, Hiker Lost in Hawaii Forest, Is Found Alive after 17 Days," *New York Times*, May 25, 2019.

190 **the anterior cingulate nucleus:** Amir-Homayoun Javadi et al., "Backtracking during Navigation Is Correlated with Enhanced Anterior Cingulate Activity and Suppression of Alpha Oscillations and the 'Default-mode' Network," *Proceedings of the Royal Society of Biological Sciences* 286, no. 1908 (2019): 20191016, doi: 10.1098/rspb.2019.1016.

191 **trying hard to navigate doesn't make us perform any better:** Heather Burte and Daniel R Montello, "How Sense-of-Direction and Learning Intentionality Relate to Spatial Knowledge Acquisition in the Environment," *Cognitive Research* 2, no. 1 (2017): 18; doi: 10.1186/s41235-017-0057-4.

APPENDIX I: WHAT TO DO IF YOU'RE LOST IN THE WILD

193 **now used worldwide as a field manual:** Robert Koester, *Lost Person Behavior* (Charlottesville, VA, dbS Productions LLC, 2008).

193 **He goes missing on a Saturday afternoon in July:** Jared Doke, "Analysis of Search Incidents and Lost Person Behavior in Yosemite National Park," Kansas University graduate degree thesis, 2012, https://kuscholarworks.ku.edu/bitstream/handle/1808/10846/Doke_ku_0099M_12509_DATA_1.pdf?sequence=1&isAllowed=y.

Index